SO HUMAN AN ANIMAL

SO HUMAN

New York

AN ANIMAL

René Dubos

MEMBER AND PROFESSOR,
THE ROCKEFELLER UNIVERSITY

CHARLES SCRIBNER'S SONS

H-8.71[C]

Printed in the United States of America

SBN 684-71753-0

Library of Congress Catalog Card Number 68-27794

To the skies of the Ile de France
and of the Hudson River Valley

FOREWORD

Each human being is unique, unprecedented, unrepeatable. The species *Homo sapiens* can be described in the lifeless words of physics and chemistry, but not the man of flesh and bone. We recognize him as a unique person by his voice, his facial expressions, and the way he walks—and even more by his creative responses to surroundings and events.

I shall discuss in the present book how each one of us has become what he is and behaves as he does, always having in mind the course of events that transformed me from a poor boy well adjusted to life in a French village into a reasonably successful citizen of New York City. A few words about my own life may explain the kind of interest I take in the personal history of my fellow men.

When I emigrated to America from my native France at the age of twenty-four, I had no plan for the future except to try my luck in the land of unlimited opportunities, but I must have been in some way preadapted to American life. The immigration officer in New York let me in with a vigorous handshake and friendly words that seemed to imply that I was the ideal type of immigrant. His warm welcome did not surprise me, because a professor from the University of Wisconsin whom I had met accidentally in Europe a few months before had encouraged me to emigrate with the remark, "I know how to pick a winner."

From the beginning, I have felt completely at ease wherever I have worked in the United States. I doubt that I could have been as healthy, successful, and happy anywhere else in the world. Yet, after more than forty years of continued residence, the most personal part of it in the Hudson River Valley, I still have some mental reservations when I say that I am an American. This is not for lack of allegiance to my adoptive country, or regret at having become an American citizen thirty years ago, but because I have not outgrown and do not want to discard the attitudes that I acquired in the small French villages where I spent my formative years and in Paris during my student days. The subtle quality of the skies, woods, and fields of the Ile de France country, and the intellectual discipline of French culture, have left an indelible stamp on my biological and mental being.

A few years after arriving in the United States, I drove from New York to the Pacific Northwest during the summer vacation. Crossing the continent over dusty corduroy roads in a second-hand car with old tires was a strenuous enterprise in the late 1920s. The most interesting experiences of the trip, however, proved to be the human contacts in the humble tourist homes and restaurants where I became acquainted with people and ways of life different from those I had known in Europe and on the East Coast.

Since I am six feet tall and have blue-green eyes, and at the time of this trip, had Viking-like flaxen hair, I looked as Nordic as the people of Anglo-Saxon and Scandinavian origin who then made up the largest percentage of the population in the Midwestern and Northwestern states. Yet it was obvious that the French culture had made me different from the tall blond young men of my age who had been born and raised in the United States. Traveling from coast to coast taught me also that, irrespective of national origin, racial type, or complexion, Westerners differed from Easterners in behavior and in tastes. Even more striking was the contrast between the American Anglo-Saxons in Seattle and the Canadian Anglo-Saxons in Vancouver.

This auto safari across the continent sharpened my awareness of the influence that geographical and social factors exert on the biological and mental being. Now that I have been around the world, I know that, contrary to what is said, even international airports have their local flavor. The ticket agents, bartenders, waitresses, salesgirls at the souvenir shops behave differently in New York from the way they do in Los Angeles or London, San Juan or Sydney. Jets and world-wide television have not altered the fact that rocky hills and alluvial plains, family farmsteads and housing developments, foster different kinds of people.

During the past forty years my professional activities as a microbiologist have given me many chances to observe in the laboratory that the characteristics of all living things are deeply affected by the conditions of their lives.

For example, I have studied how the environment in

the soil or in the mammalian body influences the multiplication and properties of microbes and how through a series of feedback processes microbial life in turn affects soil fertility and animal development. More recently, I have concerned myself with the effects that nutrition, sanitation, and housing conditions exert on the growth, resistance, life span, and behavioral characteristics of laboratory animals. The experience thus gained in the laboratory has given me the faith that scientific research, properly focused, could usefully complement the traditional humanistic approach to the study of human life.

The structure of the present book has emerged from my personal interest in man's responses to his physical and social surroundings and from my professional knowledge of the forces that can be observed in the laboratory to affect all the manifestations of life. I have used both kinds of information to illustrate that each individual is unique in the ways his innate endowment responds to his environment.

The unifying theme of this book is that all experiences leave a stamp on both physical and mental characteristics. I have placed special emphasis on the very early influences, prenatal as well as postnatal, because their effects are so profound and lasting that they have large consequences for human life. From juvenile delinquency to racial conflicts, from artistic sensibilities to national genius, few are the individual and social characteristics that are not profoundly and lastingly affected by early influences.

Since human beings are as much the product of their total environment as of their genetic endowment, it is theoretically possible to improve the lot of man on earth by manipulating the environmental factors that shape his nature and condition his destiny. In the modern world, urbanization and technology are certainly among the most important of these factors and for this reason it is deplorable that so little is done to study their effects on human life.

We claim to live in a scientific era, but the truth is that, as presently managed, the scientific enterprise is too lopsided to allow science to be of much use in the conduct of human affairs. We have accumulated an immense body of knowledge about matter, and powerful techniques to control and exploit the external world. However, we are grossly ignorant of the effects likely to result from these manipulations; we behave often as if we were the last generation to inhabit the earth.

We have acquired much information about the body machine and some skill in controlling its responses and correcting its defects. In contrast, we know almost nothing of the processes through which every man converts his innate potentialities into his individuality. Yet without this knowledge, social and technological innovations are not likely to serve worthwhile human ends.

The "square" life, as usually understood, is stifling and thwarts the responses essential for man's sanity and for the healthy development of human potentialities. All thoughtful persons worry about the future of the children who will have to spend their lives under the absurd social and environmental conditions we are thoughtlessly creating; even more disturbing is the fact that the physical and mental characteristics of mankind are being shaped *now* by dirty skies and cluttered streets, anonymous high rises and amorphous urban sprawl, social attitudes which are more concerned with things than with men.

Young people have good reason to reject the values that govern technicized societies; but protesting against conventional patterns of behavior or withdrawing from the present economic system will not suffice to change the suicidal course on which we are engaged. A constructive approach cannot be only political or social. It demands that we supplement the knowledge of things and of the body machine with a science of human life.

I have written this book in the faith that it is possible to deal scientifically with the living experience of man. The problems I have met while integrating my French heritage with the rich experience of my American life have given me the conviction that each one of us can consciously create his personality and contribute to the future, by using what the world of the present offers him to convert his hereditary and experiential past into living reality. Scientific knowledge of the effects that surroundings, events and ways of life exert on human development would give larger scope to human freedom by providing a rational basis for option and action. Man makes himself through enlightened choices that enhance his humanness.

René Dubos

CONTENTS

SO HUMAN AN ANIMAL

1. THE UNBELIEVABLE FUTURE

Rebels in Search of a Cause

This book should have been written in anger. I should be expressing in the strongest possible terms my anguish at seeing so many human and natural values spoiled or destroyed in affluent societies, as well as my indignation at the failure of the scientific community to organize a systematic effort against the desecration of life and nature. Environmental ugliness and the rape of nature can be forgiven when they result from poverty, but not when they occur in the midst of plenty and indeed are produced by wealth. The neglect of human problems by the scientific establishment might be justified if it were due to lack of resources or of methods of approach, but cannot be forgiven in a society which can always find enough money to deal with the issues that concern selfish interests.

Unfortunately, writing in anger requires talents I do not possess. This is my excuse for presenting instead a mild discussion of our collective guilt.

We claim that human relationships and communion with nature are the ultimate sources of happiness and beauty. Yet we do not hesitate to spoil our surroundings and human associations for the sake of efficiency in acquiring power and wealth. Our collective sense of guilt comes from a general awareness that our praise of human and natural values is hypocrisy as long as we practice social indifference and convert our land into a gigantic dump.

Phrases like "one world" and the "brotherhood of man" occur endlessly in conversations and official discourses at the very time that political wars and race riots are raging all over the world. Politicians and real-estate operators advocate programs for the beautification of cities and highways, while allowing the exciting grandeur of the American wilderness to degenerate into an immense ugliness. Brush is overgrowing mountain slopes that were once covered with majestic forests; industrial sewers are causing sterility in streams that used to teem with game fish; air pollutants generate opaque and irritating smogs that dull even the most brilliant and dramatic skies. The price of power, symbolized by superhighways and giant factories, is a desecration of nature and of human life.

Aggressive behavior for money or for prestige, the destruction of scenic beauty and historic landmarks, the waste of natural resources, the threats to health created by thoughtless technology—all these characteristics of our society contribute to the dehumanization of life. Society cannot be reformed by creating more wealth and power. Instead economic and technologic considerations must be made subservient to the needs, attributes, and aspirations that have been woven into the fabric of man's nature during his evolutionary and historical development.

The most hopeful sign for the future is the attempt by the rebellious young to reject our social values. Their protests indicate that mankind is becoming disturbed by increasing dehumanization and so may act in time to reverse the trend. Despite so many intellectual and ethical setbacks, despite so much evidence that human values are being spoiled or cheapened, despite the massive destruction of beauty and of natural resources, as long as there are rebels in our midst, there is reason to hope that our societies can be saved.

The social role of the rebel is symbolized by Honoré Daumier's picture *L'Emeute* (*The Uprising*) in the Phillips Memorial Gallery in Washington, D.C. The painting represents a revolutionary outbreak in nineteenth-century Paris. A handsome young man, with outstretched arms and clenched fists, is leading a crowd which appears hypnotized by his charismatic determination. His expression is intense, yet his dreamer's eyes are not focused on any particular object, person, or goal. He contemplates a distant future so indistinct that he probably could not describe the precise cause for which he and his followers are risking their lives.

Daumier's painting does not portray a particular type of rebel, or a particular cause for rebellion. Its theme is rebellious man ready to confront evil and to undertake dangerous tasks even if the goal is unclear and the rewards uncertain. The rebel is the standard-bearer of the visionaries who gradually increase man's ethical stature; because there is always evil around us, he represents one of the eternal dimensions of mankind.

The nineteenth-century rebels symbolized in Daumier's painting fought for political liberty and social equality. Today's rebels also try to identify themselves with political and social issues, such as world peace, equality of opportunities for all, or simply freedom of speech for college students. There comes to mind the caption of a popular cartoon, "Pick your own picket," which conveys with sad irony that civilized nations still have a wide range of social wrongs.

Rebellion, however, should reach beyond conventional political and social issues. Even if perfect social justice and complete freedom from want were to prevail in a world at peace, rebels would still be needed wherever the world is out of joint, which now means everywhere. Rebellion permeates all aspects of human life. It originates from the subconscious will of mankind not to surrender to destructive forces. But rebelling is not the same as defining a cause that would improve the quality of human life, or formulating a constructive program of action. Marching in a parade is easier than blazing a trail through a forest or creating a new Jerusalem. Daumier's hero looks like many rebels in our midst. He is fighting against evil rather than for a well-defined cause. Like most of us, he is a rebel without a program.

Our society is highly expert in controlling the external world and even the human mind, but our relationships with other human beings and the rest of creation are constantly diminishing in significance. This society has more comfort, safety, and power than any before it, but the quality of life is cheapened by the physical and emotional junk heap we have created. We know that life is being damaged by the present social conditions, but we participate nevertheless in a system that spoils both the earth and human relationships. Most contemporary rebels, like the rest of us, are unwilling to give up the personal advantages so readily derived from the conditions we all know to be objectionable. Nevertheless, rebels play a useful social role; at least they voice our collective concern and make us aware of our collective guilt. But the acknowledgment of guilt is not enough.

Rumblings against the present state of things remain amorphous and ineffective largely because existing trends, customs, and policies cannot be changed merely by negative acts. Positive beliefs are required. Alternatives will not emerge through piecemeal evolution; their development demands an intellectual and emotional revolution. We cannot transform the world until we eliminate from our collective mind the concept that man's goals are the conquest of nature and the subjection of the human mind. Such a change in attitude will not be easy. The search for the mastery of nature and for unlimited growth generates a highly stimulating, almost intoxicating atmosphere, whereas the very hint of approaching stabilization creates apathy. For this reason, we can change our ways only if we adopt a new social ethic—almost a new social religion. Whatever form this religion takes, it will have to be based on harmony with nature as well as man, instead of the drive for mastery.

We have already accepted in principle, even though we rarely put into practice, the concept of human brotherhood. We must now take to heart the biblical teaching, "The Lord God took the man and put him into the Garden of Eden to dress it and to tend it" (Genesis 2:15). This means not only that the earth has been given to us for our enjoyment, but also that it has been entrusted to our care.

Technicized societies thus far have exploited the earth; we must reverse this trend and learn to take care of it with love.

On the occasion of the annual meeting of the American Association for the Advancement of Science in 1966, the American historian Lynn White, Jr., pleaded for a new attitude toward man's nature and destiny. He saw as the only hope for the world's salvation the profoundly religious sense that the thirteenth-century Franciscans had for the spiritual and physical interdependence of all parts of nature. Scientists, and especially ecologists, he urged, should take as their patron Saint Francis of Assisi (1182–1226).[1] But was not Francis one of the rebellious youths of his time—before the Church recognized that he was serving God by reidentifying man with nature? Francis, like Buddha, spent his early years in ease and luxury but rejected bourgeois comforts in search of more fundamental values. The contemporaries of both probably regarded them as beatniks.

The name Saint Francis and the word ecology are identified with an attitude toward science, technology, and life very different from that which identified man's future with his ability to dominate the cosmos. The creation of an environment in which scientific technology renders man completely independent of natural forces calls to mind a dismal future in which man will be served by robots and thereby himself become a robot. The humanness of life depends above all on the quality of man's relationships to the rest of creation—to the winds and the stars, to the flowers and the beasts, to smiling and weeping humanity.

Shortly before his death in 1963, the English novelist and essayist Aldous Huxley lamented on several occasions the fact that literature and the arts have not derived any worthwhile inspiration from modern science and technology. He thought the reason for this failure was that writers and artists are unaware of modern scientific and technological developments.[2] This may be part of the explanation but only a very small part. Like most other human beings, writers and artists are primarily concerned with perceptions, emotions, and values which the scientific enterprise must deliberately ignore. Yet scientists should not be satisfied with studying the biological machine whose body and mind can be altered and controlled by drugs and mechanical gadgets. They should become more vitally concerned about the nature and purpose of man. Only thus can they learn to speak to man not in a specialist's jargon but in a truly human language.

The New Pessimism

As the year 2000 approaches, an epidemic of sinister predictions is spreading all over the world, as happened among Christians during the period preceding the year 1000. Throughout the tenth century, Norsemen and Saracens incessantly raided Western Europe, disorganizing daily life and secular institutions, pillaging churches and monasteries. The rumor spread that the year 1000 would mark the end of the world and that a new spiritual universe would come into existence. Even those who did not believe that the world would come to an end probably assumed that living conditions would be corrupted by the barbaric invaders.

Prophets of gloom now predict that mankind is on a course of self-destruction, or that, in the unlikely event of its survival, it will progressively abandon the values and amenities of Western civilization. Nuclear warfare, environmental pollution, power blackouts, the progressive erosion of public services constitute direct and obvious threats to human existence. Furthermore, social regimentation and loss of privacy may soon reach levels incompatible with the traditional ways of civilized life. The established order of things appears to be threatened by technological and social forces that increasingly dominate the world, just as it was threatened by the raiding Norsemen and Saracens ten centuries ago.

Many observers of the contemporary scene would agree with the following words by the American journalist James Reston in the most influential daily newspaper of the most prosperous city in the world: "The old optimistic illusion that we can do anything we want is giving way to doubt, even to a new pessimism."[3] Newspaper headlines daily seem to confirm the belief that the problems of the cities, the races, and the nations are beyond our control.

Apprehension is most widespread and expresses itself most clearly with regard to nuclear warfare, threats to health, the rise of automation, and other ill-defined consequences of scientific technology. Popular articles entitled "The Truth About . . ." almost uniformly refer to the dangers of technological or medical innovations. The new pessimism, however, has other determinants which transcend the fear of annihilation and affect the quality of life. In particular, science is being accused of destroying religious and philosophic values without substituting other guides to behavior or providing a meaningful picture of the universe. The disintegrating effect of loss of belief was pungently expressed a generation ago by the American philosopher John Dewey in his warning that a culture which permits science to destroy traditional values, but which distrusts its power to create new ones, is destroying itself. Man finds it difficult to live without ultimate concern and faith in the significance of his destiny.

The malaise has now extended to the scientific community itself. While all scientists still believe that the opportunities for the extension of knowledge are boundless, many are beginning to doubt the wisdom and safety of extending much further some of the applications of knowledge.[4] In addition, there have been claims that limitations inherent in the very structure of the physical world may soon slow down, then interrupt altogether, the development of the scientific technologies which have resulted in the most spectacular achievements of our age. Airplanes cannot practically fly much faster than at the present supersonic speeds; electronic computers are approaching theoretical limits of speed and efficiency; high-energy accelerators cannot long continue to become larger and more powerful; even space travel will have achieved its human possibilities within a very few decades.[5]

The most important factor in dampening the euphoria that until recently was universal in scientific circles is the social and economic necessity of imposing directions and limitations to many technological developments. The current discussions concerning the advisability of devoting large resources to the manned space program have brought to light difficulties in reconciling the demands of certain technologies with more traditional human needs.

A few years ago, American scientists could state, "We *must* go to the moon, for the simple reason that we *can* do it"—echoing President John F. Kennedy, who in turn had echoed the statement by the English mountain climber George Mallory that Mount Everest *had* to be climbed, simply because it was there.[6] Such statements are admirable to the extent that they express man's determination to accept difficult challenges, whenever and wherever there is some chance that the effort will lead to spectacular feats. But dashing expressions do not constitute an adequate substitute for the responsibility of making value judgments.

There are many good scientific reasons for accepting the staggering human, financial, and technological effort required to explore space and to land a man on the moon. There are equally good reasons, however, for undertaking other kinds of difficult and challenging tasks—such as exploring the earth itself or the depths of the oceans, probing into the nature of matter and energy, searching for the origins of man and his civilizations, controlling organic and mental disease, striving for world peace, eliminating city slums, preventing further desecration of nature, or dedicating ourselves to works of beauty and to the establishment of an harmonious equilibrium between man and the rest of creation.

Laymen as well as scholars can think of many projects at least as important and interesting as space travel or lunar exploration, and just as likely to succeed. But limitations of resources make it impossible to prosecute all worthwhile projects at the same time. Hence, the statement that we *must* do something because we *can* do it is operationally and ethically meaningless; it is tantamount to an intellectual abdication. Like other responsible human beings, scientists and sociologists must discriminate; their choice of goals must be made on the basis of value judgments.

The problem of choice is greatly complicated by the fact that technological advances endlessly create new dilemmas, since every innovation has unforeseen consequences. Social regimentation, traffic jams, environmental pollution, constant exposure to noise and other unwanted stimuli are but a few of the undesirable accompaniments of economic and technological growth. Indeed, many innovations that have enhanced the wealth and power of our society in the past threaten to paralyze it at a later date. Abundance of goods, excess of comfort, multiplicity of means of communication are generating in the modern world situations almost as distressing as the ones that used to result from shortages of food, painful physical labor, and social isolation. We are creating new problems in the very process of solving those which plagued mankind in the past.

During recent years experts in the natural and social sciences have repeatedly pointed out that the erratic and misguided growth of technology and urban conditions now poses as serious a threat as the undisciplined growth of the world population. Economic affluence and scientific breakthroughs appear paradoxically to remove man still further from the golden age.[7]

The new pessimism derives in large part probably from the public's disenchantment at the realization that science cannot solve all human problems. Furthermore, the public is beginning to realize that whenever scientists make claims for support of their activities in the name of relevance to industrial technology, they are in fact making value judgments concerning the importance of technology in human life, judgments for which they have no special competence. A few spectacular technological failures might suffice to generate a bankruptcy of science.

Phrases such as the classical age, the age of faith, the age of reason, or the romantic age may not correspond to historical realities, but they convey nevertheless mankind's nostalgic longing for certain qualities of life that most people, rightly or wrongly, associate with the past. In contrast, we prosaically designate our own times the atomic age, space age, age of automation, antibiotic age—in other words, the age of one or another technology. These terms are used approvingly by technologists and disparagingly by humanists. The one term which has received almost universal acceptance is age of anxiety.

Social and technological achievements have spread economic affluence, increased comfort, accelerated transportation, and controlled certain forms of disease. But the material satisfactions thus made possible have not added much to happiness or to the significance of life. Not even the medical sciences have fulfilled their promises. While they have done much in the prevention and treatment of a few specific diseases, they have so far failed to increase true longevity or to create positive health. The age of affluence, technological marvels, and medical miracles is paradoxically the age of chronic ailments, of anxiety, and even of despair. Existentialist nausea has found its home in the most affluent and technologically advanced parts of the world.

Present-day societies abound in distressing problems, such as racial conflicts, economic poverty, emotional solitude, urban ugliness, injustice in all its forms, and the collective lunacy that creates the threat of nuclear warfare. But modern anxiety has deeper roots that reach into the very substance of each person's individuality. The most poignant problem of modern life is probably man's feeling that life has lost significance. The ancient religious and social creeds are being eroded by scientific knowledge and by the absurdity of world events. As a result, the expression "God is dead" is widely used in both theological and secular circles. Since the concept of God symbolized the totality of creation, man now remains without anchor. Those who affirm the death of God imply thereby the death of traditional man—whose life derived significance from his relation to the rest of the cosmos. The search for significance, the formulation of new meanings for the words God and Man, may be the most worthwhile pursuit in the age of anxiety and alienation.

Alienation is a vague word, but it denotes an attitude extremely widespread at present in affluent societies. Feeling alienated is an ancient experience which has taken different forms in the course of history. Many in the past experienced forlornness because the cosmos and the human condition appeared to them meaningless and pointless. Jean Jacques Rousseau, in the eighteenth century, traced alienation to the estrangement of man from nature that in his view resulted from artificial city life. Karl Marx, in the nineteenth, coined the word *Entfremdung*—rendered as "alienation" by his translators—to denote both the plight of the industrial worker deprived of the fruits of his production, and the depersonalization of labor in mechanized industries.

Many forms of alienation now coexist in our communities. The social and cultural malaise affects not only disenchanted intellectuals, industrial labor, and the poor classes, but also all those who feel depersonalized because circumstances compel them to accept mass standards which give them little chance to affirm their identity. Alienation is generated, furthermore, by the complete failure of even the most affluent societies to achieve harmonious relationships between human life and the total environment. The view that the modern world is absurd is no longer limited to the philosophical or literary avant-garde. It is spreading to all social and economic groups and affects all manifestations of life.

Psychologists, sociologists, and moralists tend to attribute anxiety and despair to the breakdown of intimate social relationships, with the attendant personal loneliness so pervasive in modern cities. The breakdown, however, is not limited to the interplay between human beings. It extends to the interplay between man and the natural forces that have shaped his physiological and mental self and to which his most fundamental processes are still bound. Chaos in human relationships has the same origin as chaos in the relationships between man and his environment.

In all countries of Western civilization, the largest part of life is now spent in an environment conditioned and often entirely created by technology. Thus one of the most significant and disturbing aspects of modern life is that man's contacts with the rest of creation are almost always distorted by artificial means, even though his senses and fundamental perceptions have remained the same since the Stone Age. Modern man is anxious, even during peace and in the midst of economic affluence, because the technological world that constitutes his immediate environment, by separating him from the natural world under which he evolved, fails to satisfy certain of his unchangeable needs. In many respects, modern man is like a wild animal spending its life in a zoo; like the animal, he is fed abundantly and protected from inclemencies but deprived of the natural stimuli essential for many functions of his body and his mind. Man is alienated not only from other men, not only from nature, but more importantly from the deepest layers of his fundamental self.

The aspect of the new pessimism most commonly expressed is probably the belief that decrease in individual freedom is likely to result from increasing densities of population and the consequent need to accept a completely technicized urban environment. A heavy and repetitious anthology could be composed of writings by all kinds of scholars lamenting the sacrifice of personality and freedom at the altar of technological regimentation. As society becomes ever more highly organized, the individual will progressively vanish into the anonymous mass.

In his book *The Myth of the Machine*, the American critic Lewis Mumford predicts a future in which man will become passive and purposeless—a machine-conditioned animal designed and controlled for the benefit of depersonalized, collective organizations.[8]

Former Secretary of State Dean Acheson expresses a similar concern in his recently published memoirs. He puts the golden age of childhood "quite accurately between the last decades of the nineteenth century and the first half of the 1920s . . . before the plunge into a motor age and city life swept away the freedom of children and dogs, put them both on leashes and made them the organized prisoners of an adult world."[9]

The new pessimism considers it almost inevitable that the complexity of social structures will result in social regimentation and that freedom and privacy will come to be regarded as antisocial luxuries. Under these conditions, the types of men more likely to prosper will be those willing to accept a sheltered but regimented way of life in a teeming and polluted environment from which both wilderness and fantasy will have disappeared. The world may escape catastrophic destruction, but if present trends continue our descendants will find it difficult to prevent a progressive decadence of the social order of things. The tide of events will bring about simultaneously, paradoxical as it may seem, the fragmentation of the person and the collectivization of the masses.

Naturally there are some optimists among the modern soothsayers, but the new Jerusalem they envision is little more than a dismal and grotesque magnification of the present state of affairs. They predict for America a gross national product of many trillions of dollars and an average family income so large that every home will be equipped with more and more power equipment and an endless variety of electronic gadgets. Drugs to control the operations of the body and the mind, complicated surgery, and organ transplantations will make it commonplace to convert ordinary citizens into "opti-men." The working day will be so short and the life span so long that countless hours will be available for the pursuit of entertainment—and eventually perhaps merely for the search for a *raison d'etre.*

Most modern prophets, in and out of the academies, seem to be satisfied with describing a world in which everything will move faster, grow larger, be mechanized, bacteriologically sterile, and emotionally safe. No hand will touch food in the automated kitchen; the care and behavior of children will be monitored electronically; there will be no need to call on one's friends because it will be possible to summon their voices, gestures, shapes, and complexions on the television-paneled walls of the living room; life will be effortless and without stress in air-conditioned houses romantically or excitingly lighted in all sorts of hues according to one's moods; exotic experiences will be safely and comfortably available in patios where artificial insect sounds and the proper degree of heat and moisture will create at will the atmosphere of a tropical night or a New England summer day. Actual examples of the dismal life that technological prophets envision for the future can be found increasingly in periodicals and books, notably the recently published *The Year 2000. . . .*[10]

Admittedly, modern prophets also have visions reaching beyond the mere provision of effortless comfort and entertainment. But what they then imagine has the absurd quality of supersonic planes so rapid that travelers are back where they started from before having finished their first cocktail. As to the prophecies concerning space travel, or life on the bottom of the oceans, their chief purpose would seem to be to provide images of new environments in which boy meets girl and where good guy overcomes bad guy—in other words, where human beings behave exactly as they do on earth. Young people in lovers' lanes and desperadoes with their big hats and guns seem to be as essential to the unbelievable world of the future as to the old-fashioned dime novels and Western scenarios.

Science-fiction writers, abetted by not a few distinguished scientists, have indulged in a game of overpromise, which will inevitably lead to a letdown when it becomes obvious that the promises cannot be fulfilled. There exists in the modern world a pathological trend to view man's future from Mount Olympus, assiduously averting the eyes from the valleys of want, sorrow, and tears. Oddly enough, the natural sciences today provide the easiest and cheapest roads of escape from reality.

The word "unbelievable" (or its equivalent "incredible") is ambiguous. As commonly used, it denotes events or situations so extraordinary that they are difficult to believe but are nevertheless true. The present writings about all the marvels of the "unbelievable future" intend to convey this sense of impossible but true. Etymologically, however, unbelievable has a much more negative meaning, and it is in the sense of actual impossibility that the word will be used here. The "opti-man" imagined by the prophets of dismal optimism turns out to be not only a hollow man, but also a pseudo-man; not only would he be devoid of the attributes that have given its unique value to the human condition, but he would not long survive because he would be deprived of the stimuli required for physical and mental sanity.

The kind of life so widely predicted for the twenty-first century is unbelievable in the etymological sense because it is incompatible with the fundamental needs of man's nature. These needs have not changed significantly since the Late Stone Age and they will not change in the predictable future; they define the limits beyond which any prediction of the future becomes literally unbelievable.

Whatever scientific technology may create, *l'homme moyen sensuel* will continue to live by his senses and to perceive the world through them. As a result, he will eventually reject excessive abstraction and mechanization in order to reestablish direct contact with the natural forces from which he derives the awareness of his own existence and to which he owes his very sense of being.

The one possible aspect of the future seldom discussed by those who try to imagine the world-to-be is that human beings will become bored with automated kitchens, high-speed travel, and the monitoring of human contacts through electronic gadgets. People of the year 2000 might make nonsense of the predictions now being published in the proceedings of learned academies and in better-life magazines, simply by deciding that they want to regain direct contact with the natural forces that have shaped man's biological and mental being. The visceral determinants of life are so permanent, and so demanding, that mankind cannot long safely ignore them. In my opinion, the world in the year 2000 will reflect less the projections of technologists, sociologists, and economists than the vital needs and urges of biological man.

At the end of his *Education*, written in 1905, the American historian Henry Adams gloomily predicted that the cult of the Dynamo was to be the modern substitute for the cult of the Virgin.[11] The present scene appears to confirm his prediction, but the future may still prove him wrong. One begins to perceive disenchantment among the worshipers of the Dynamo, and, more importantly, there are encouraging signs of unrest in the younger generation. Frequently in the past the son rejected what the father had taken for granted, and civilization thus took a step forward. Beatniks, hipsters, teddy boys, provos, hooligans, *blousons noirs,* and the countless other types of rebellious youths are probably as ignorant, foolish, and irresponsible as conventional people believe them to be. But conventionality rarely has the knack of guessing who will shape the future. The substantial citizen of Imperial Rome and the orthodox Jews of the synagogue looked down on the small tradesmen, fishermen, beggars, and the prostitutes who followed Jesus as he preached contempt for the existing order of things. Yet Imperial Rome and the Temple collapsed, while Jesus' followers changed the course of history.

The vision of the future, as seen in the light of the new intellectual pessimism or of the dismal optimism of some technologists, would be terribly depressing if it were not for the fact that it so much resembles visions of the future throughout history. Pessimists have repeatedly predicted the end of the world, and utopians have tried to force mankind into many forms of straitjackets. Those who did not live before 1789, wrote the French statesman Talleyrand, have not known the *douceur de vivre*. This melancholy belief did not prevent Talleyrand from living very successfully to the age of eighty-four. When he died in 1838, the storming of the Bastille was half a century past. The Industrial Revolution had begun, and the world certainly looked uncouth and dark to many genteel souls. But we now realize that it was a beginning rather than an end. Like Talleyrand, and like society after the Industrial Revolution, we too shall probably manage to find an acceptable formula for our times. Fortunately, the creativeness of life always transcends the imaginings of scholars, technologists, and science-fiction writers.

Toward a New Optimism

Despite the forebodings of the tenth century, the world did not come to an end in the year 1000, nor did Europe take to barbaric ways of life. The Saracens assimilated Greek learning and transmitted it to the Mediterranean universities, from which it spread all over the Western world. The Norsemen became Christianized, and, far from destroying civilization, their uncouth barons created monasteries, churches, cathedrals, and town halls almost as fast as they built fortified castles. The Norman rulers spread first Romanesque and then Gothic architecture all over Europe to honor the Virgin and the saints; in many places also their courts provided a chivalrous atmosphere in which the troubadours converted the worship of the Virgin into the cult of womanhood.

If the rebellious young succeed in discovering a formula of life as attractive as that of the troubadours, we may witness in the twenty-first century a new departure in civilization as occurred in Europe after it recovered from the fears of the tenth century. To be humanly successful, the new ages will have to overcome the present intoxication with the use of power for the conquest of the cosmos, and to rise above the simple-minded and degrading concept of man as a machine. The first move toward a richer and more human philosophy of life should be to rediscover man's partnership with nature.

The undisciplined and incoherent expansion experienced by technicized societies during the past few decades would certainly spell the end of the human condition if it were to continue much longer. Doing more and more of the same, at a faster and faster pace, contributes neither to happiness nor to the significance of life. In the past, great prosperity has often damaged human values and generated boredom; the environmental crisis in the modern world indicates, furthermore, that mismanaged prosperity may destroy human life altogether.

The fact that economic affluence commonly leads to absurdity had been noted in the nineteenth century by Ralph Waldo Emerson, who predicted in his journal that American prosperity "would go on to madness." Many now believe that modern life in large cities is dangerously close to the state of madness!

Emerson was far-sighted when he wrote these words, because until the 1940s political reformers, economists, technologists, and scientists had no reason to question the usefulness of their efforts. Social action based on objective knowledge then appeared unequivocally beneficial; it commonly resulted in greater personal freedom, made life safer and more comfortable, protected man from irrational fears by enlarging his view of the cosmos and his understanding of biological processes. Until our times, moving onward with scientific technology could be legitimately identified not only with the creation of wealth, but also with social progress.

North America provided an ideal setting for the euphoria of the nineteenth and early twentieth centuries, which was based on the belief that industrial civilization would inevitably generate happiness by increasing comfort and creating more, better, and cheaper goods. The vastness and emptiness of the continent made it easy for the settlers to accept the myth of the ever-expanding frontier. The mood of optimistic nomadism which has been so influential in shaping American attitudes and institutions certainly derived in large part from the nineteenth-century faith that one could always move on to greener pastures. After the whole continent had been occupied, the explosive development of science and technology provided grounds for even greater optimism by opening exciting vistas for knowledge and for technological enterprise and thus providing new, apparently endless frontiers for economic growth.

In his unfinished epic poem, *Western Star*, published in 1943, the American poet Stephen Vincent Benét tried to convey the overpowering urge to move on and on without much concern for the point of destination.[12] He regarded progress *per se*—moving forward—as characteristic of the American genius. The spirit of the westward movement came through in the lines

> We don't know where we're going,
> but we're on our way.

or again,

> Americans, who whistle as you go!
> (And where it is you do not really know,
> You do not really care.)

Stephen Vincent Benét had hoped that *Western Star* would constitute a sort of *Odyssey.* He failed because his literary talents were not equal to the task, but his poem is of historical importance nevertheless on two different accounts. For one thing, it clearly expresses, even though in a pedestrian manner, the euphoric urge for expansion which characterized the nineteenth and twentieth centuries. The other historical interest of the poem is that the date of its publication almost coincided with the end of the era which valued expansion for its own sake. Since 1950, the urge for economic growth has been increasingly overshadowed by public concern with the undesirable consequences of growth: crowding, environmental pollution, traffic jams, surfeit of goods, and all the other nauseating and catastrophic by-products of excessive population, production, and consumption. Men of the twentieth century may still be whistling on their way, but deep in their hearts they are worrying about where they should go. Often they are not even sure whether they should keep on going or try to retrace their steps.

There are several reasons for the widespread skepticism concerning the advantages and even the possibility of unlimited technological growth. One is the awareness already mentioned that beyond a certain point prosperity and abundance of goods become meaningless. It is increasingly apparent, furthermore, that certain present trends are self-limiting because they lead to absurdities which, if continued, generate countertrends. The growing interest in crafts, home cooking, folk dances, and the various forms of "be-ins" certainly represents a trend against the standardization of industrial goods and commercial entertainment. The flow of population from the heart of the city to suburbia, then to exurbia, and then back to the city may be another example of trend-countertrend.

The view that Western civilization must abandon its growth myth should not be confused with the thesis of the German philosopher Oswald Spengler in *The Decline of the West* (1918) that the Western world cannot escape decadence. Rather it constitutes an expression of my faith that Western culture, and especially Western science, can be rededicated to values more lasting and more significant than those heretofore identified with technological and economic growth. The new optimism finds its sustenance in the belief that science, technology, and social organization can be made to serve the fundamental needs and urges of mankind, instead of being allowed to distort human life.

There is fear of science among the general public and resentment against it among classical scholars. But stronger and more widespread than this hostility is the belief that scientific techniques will be needed to solve the world's problems, including those created by scientific technology itself. Witness the insistent demands from the executive branch of the government, from Congressional committees, and from various private organizations that scientists direct their efforts more pointedly to the problems of man in the modern world. This subject is discussed at some length in later chapters; suffice it here to outline in very general terms some contributions that science could make to the new humanism.

Ever since the seventeenth century, science has been concerned primarily with atomistic descriptions of substances and phenomena. Its philosophical heroes have been Democritus (fifth century B.C.) and René Descartes (1596–1650), both of whom taught that the way to knowledge is to separate substances and events into their ultimate components and reactions. The most pressing problems of humanity, however, involve relationships, communications, changes of trends—in other words, situations in which systems must be studied as a whole in all the complexity of their interactions. This is particularly true of human life. When life is considered only in its specialized functions, the outcome is a world emptied of meaning. To be fully relevant to life, science must deal with the responses of the total organism to the total environment. An earlier Greek philosopher, Heraclitus, who taught that everything is flux, may well replace Democritus as the precursor of the new scientific humanism.

We are worried by the universal threat to natural resources and shudder at the raping of nature caused by scientific technology and overpopulation. We wonder, indeed, whether man can long survive in the artificial environment he is creating. To approach these problems constructively, we must learn more of the complex interplays between man, his technologies, and his environment. We must define with greater precision the determinants of man's responses to environmental forces—his innate limitations as well as his potentialities, his acquired characteristics as well as his aspirations. A sophisticated form of ecology will have to complement Democritus' atomicism and Descartes' reductionism.

We lament the dehumanization of man. Anthropology has taught us that man acquired his humanness while evolving in intimate relation with other living things and we know that all phases of his development are still conditioned by the social stimuli that he receives in the course of his life. We must develop a science of modern man considered not as an object, but rather in his interplay with other human beings—during both the emotional depth of individual encounters and the less demanding ordinary social relationships.

Since science and the technologies derived from it are now playing such an immense role in human societies and changing them so rapidly, man can survive only by continuously adjusting himself to ever-new conditions. Such adaptive processes are the inevitable consequences of social and technological innovations. Many modern thinkers—biologists as well as technologists, religious believers as well as atheists—go so far as to state that man's increasing dependence on the machine constitutes an essential fact in his evolution—the process that the Jesuit archaeologist Pierre Teilhard de Chardin has termed hominization.[13] Despite the authority this concept has thus received from theologians, philosophers, scientists, and enlightened laymen, the view that man's future is linked to technology can become dangerous if accepted uncritically. Any discussion of the future must take into account the inexorable biological limitations of *Homo sapiens.*

Acknowledgment of these limitations need not imply either a static view of man's nature or a resigned acceptance of the status quo for the human condition. Looking by night on the towering black mass of Chicago's buildings, the American architect Frank Lloyd Wright came as early as 1901 to the conclusion that "if this power must be uprooted that civilization may live, then civilization is already doomed."[14] Scientific technology cannot and should not be uprooted; not only has it become indispensable for man's survival but it has enriched his perceptions, enlarged his vision, and deepened his concept of reality. To a very large extent the continued unfolding of civilization will depend on the imaginative creativity of scientific technologists. But it would be dangerous to assume that mankind can safely adjust to all forms of technological development. In the final analysis, the frontiers of social and technological innovations will be determined not by the extent to which man can manipulate the external world but by the limitations of his own biological and emotional nature.

Total wisdom requires the attitudes of both the sage and the scientist, integrated on the high ground of man's meta-technical being and destination. Pure rationalism degrades wisdom to the level of bloodless abstractions, and technocratic thinking reduces man to a machine. On the other hand, science in all its forms and applications is now creating values that transcend the mechanical aspects of life. A social philosophy suitable for our times must therefore include scientific humanism. Our formidable knowledge of physical forces and inanimate matter must be supplemented by scientific knowledge of the living experience and by awareness of human aspirations.

In his Sigma Xi–Phi Beta Kappa address before the annual meeting of the American Association for the Advancement of Science in 1966, the American astronomer W. O. Roberts dared to raise questions about the nature and purpose of man and about what constitutes a good life and a good society. A generation ago such questions would have been considered outside the province of scientists. But Dr. Roberts was expressing the universal uneasiness about the future when he pleaded for a concern with ultimate purpose in human life and for a philosophy geared to the chain-reacting growth of science. He referred to science as "a wellspring of our discontent," not because of its obvious influence on the practical aspects of our day-to-day life, but because of its impact on man's changing conception of himself and his world.[15]

In addition to the science of material things we must develop a science of humanity. Both together will constitute the humanism of the future, a new kind of *Gai Savoir*.[16]

2.

MAN'S NATURE AND HUMAN HISTORY

The Humanness of Prehistoric Man

In a cave of eastern Oregon, near the village of Fort Rock, the American archaeologist L. S. Cressman and his students discovered in 1936 a cache of seventy-five sandals buried in volcanic pumice. Most of these are now on display in the archaeological museum of Oregon State University at Eugene. The sandals are woven from shredded sagebrush bark twisted into tight ropes thinner than an ordinary pencil, and they exhibit great uniformity in workmanship. They measure from 9 to 12 inches in length and would therefore fit a modern man. Yet archaeological evidence, recently confirmed by carbon 14 dating, proves that they were manufactured by Indians some 9,000 years ago.[1]

Eastern Oregon is now a desert country, but during the Late Paleolithic period there was a large lake in the region where the sandals were discovered. The Indians who then lived around the lake had apparently developed a complex social organization, as is attested by the storage of so many artifacts in a single cave. Human occupation of the area was probably interrupted by the volcanic eruption that deposited the layer of pumice over the floor of the Fort Rock cave. On first contact, the hot pumice charred the sandals somewhat, but after cooling it acted as a protective layer preventing further deterioration from the inclemencies of the weather and attack by microbes and insects.

The detailed description of the sandals published by Dr. Cressman gives an idea of the workmanship of the Stone Age Indians:

"Five pieces of rope laid lengthwise to the long axis of the foot served as warps and were fastened tightly together by twining weft strands. The toe ends of the warp strands, left untwisted, were folded back over the toes to form a protective pocket and, held slackly together, were fastened to the sides of the sandals by rather loose twining. The tie-string was drawn through a series of loops around the heel and on the sides made by the looped weft strands. All were alike except a few which had cords running tightly and slightly diagonally across the sole. The purpose of this cord is unknown; if it had been serviceable as a non-skid device, it would probably have been more widely used. In some sandals pine needles had been added for padding."[2]

Other types of sandals woven by Indians many thousand years ago have been discovered in several parts of the United States and can be seen in anthropological museums. They differ in workmanship and in style from region to region, indeed from one cave to the other in the same region. Sandals found in Catlow Cave, Oregon, are not as well made as the ones from Fort Rock, even though the two sites are close to each other and were occupied by Indians during approximately the same period.

Sandals discovered in still other caves are made of tule, or more rarely of grass, instead of sagebrush bark. Irrespective of the material of which they are made, some are obviously designed for rough usage while others are more refined in style—some so elegant in design and workmanship that they would not seem out of place in a New York Fifth Avenue shop.

A few of the ancient Indian sandals are child size and have rabbit fur woven into them as if for warmth and softness. Again quoting Dr. Cressman, an Oregon cave yielded ". . . a pair of small sandals for a five- or six-year-old child, tied together by strings as we might tie a pair of sneakers. Nearby were two toy baskets, and a little farther off was a dart for a 'dart and wheel' game. Because these objects lay close together, we are sure that they were the sandals and playthings of an Indian girl who had lived in that cave several thousand years ago. One day something happened; sandals and toys were left where they were last used, as we might leave shoes and toys on the living room floor."[3]

The prehistoric sandals, large or small, crude or stylish, create a sense of kinship with the human beings who made and used them many thousands of years ago. The variety of workmanship and the design for various types of usage make it apparent, better than words ever could, that Stone Age man had mastered many skills and developed a complex familial and societal organization.

The humanness of prehistoric man is of direct relevance

to our own lives, because we have inherited from him most of our physiological and mental characteristics and we share with him the same fundamental needs and urges. Many aspects of modern life are profoundly affected by the forces that shaped *Homo sapiens* and his life as far back as the Late Paleolithic or Old Stone Age, more than 100,000 years ago.

Homo sapiens does not differ from animals so much by his ability to learn as by the kinds of things he learns, in particular by the accumulation of his social experiences in the course of collective enterprises over thousands of generations. In other words, the human species is best characterized by its social history.[4]

As conventionally defined, history begins with the period of the oldest written documents that have come down to us; these date from the Sumerian civilization about 6000 years ago. However, so many well-preserved artifacts providing precise information on human life have survived from the Stone Age that the preliterate period can also be included in the sociobiological history of mankind.

As far as is known, only man can control and use fire. This first technological achievement of mankind, which occurred perhaps a million years ago, is celebrated in the legend of Prometheus. The drama of the legend, and the role of hearth and altar in human traditions, suggest that man recognized very early that the mastery of fire has played a large role in his emergence from brutish life. For approximately 100,000 years, human life has been identified not only with the use of fire, but also with shelter, clothing, tools, weapons, complex social structures, and with the practice of some form of magic or worship. The fact that burial was practiced during the Stone Age, even by Neanderthal man, suggests some form of ultimate concern.

The distribution of human skeletons and artifacts in Europe indicates that early man moved slowly north or south concomitantly with the retreats or advances of the Pleistocene ice sheets. Such migrations, extended over thousands of years, were probably at first largely unconscious. Since man in the Old Stone Age was primarily a hunter, he did not follow the movements of the ice *per se*, but rather the animals on which he depended for his livelihood. By the fourth ice age, however, large numbers of men seem to have remained in frigid Europe. Presumably their technological and social culture was then sufficiently advanced for them to develop the practices required for survival under difficult climatic conditions; it is possible also that large numbers of game animals persisted nearby. This was approximately 30,000 to 50,000 years ago; from that time on, human cultures rapidly became more complex and man progressively achieved increasing control over his environment.

By the end of the Old Stone Age, man's kit of artifacts included needles, scrapers, knives, harpoons, spearheads, engraving tools, and a host of other objects. Certain flat stone containers apparently served to hold the burning fat or oil used to light the caves, probably during rituals and for painting. The famous statuettes of women, now referred to as Paleolithic Venuses, and the spectacular cave paintings and drawings of Europe attest not only to the artistic abilities of prehistoric man but also to the existence of elaborate beliefs and magical practices. In its magical forms at least, religion seems to have emerged simultaneously with humanness.

Paleolithic weapons, tools, sculptures, drawings, and paintings, like the sandals of the Oregon Indians, differed in workmanship and style from one area to another. There were regional styles even then, and for all we know some of these may be reflected in human life even today. But despite local differences, the ways of life throughout the inhabited world probably had many characteristics in common and remained much the same as long as man continued to derive his livelihood chiefly from hunting. The situation changed when the ice made its final retreat from Europe, and when human populations became less nomadic. Civilization then entered the era generally referred to as the Mesolithic.

The domestication of the dog and the manufacture of bows and arrows are among the Mesolithic innovations. There is also some evidence that pottery manufacture was invented by Mesolithic man in northern Europe independently of its development in the Near East. However, of greatest importance for the future was the differentiation in ways of life that resulted from man's becoming progressively specialized in the utilization of the plant and animal resources peculiar to different regions. Some human groups adapted their life to grassland hunting; others to hunting and fishing in the deep forest; still others to shellfish collecting on the coast or to the exploitation of other marine resources. The tendency of Mesolithic life to achieve a close partnership with the local environment probably resulted in greater geographical stability of the populations, and in the progressive emergence of various human types. These changes prepared man for the Neolithic phase of civilization.

Neolithic life is identified with the manufacture of polished stone tools and weapons, the rapid development of the potter's art, and the domestication of plants and animals.

Agricultural techniques not only increased food resources but made them more dependable; this resulted in a rapid increase in population and greater stability of human settlements.

In the Old World, the change from the hunting to the agricultural way of life apparently occurred independently in two separate areas, probably first in southwestern Asia, then in southeastern Asia. From its areas of origin, approximately 10,000 years ago, the Neolithic pattern of life diffused rapidly into Europe, Africa, and the rest of Asia, taking very different forms under the influence of local conditions and resources. Wheat, barley, and rye, cattle, horses, sheep, and goats soon became common in most of the human settlements identified with the cultures of southwestern Asia. Rice, sweet potatoes and other root crops, chickens, and pigs were more characteristic of life in southeastern Asia.

The greater abundance and dependability of food supplies, with the attendant stability of human settlements, set the stage for the rise of the great civilizations which are more or less identified with the Bronze Age. Men of the Near East learned to alloy copper and tin sometime around 5,000 B.C.; they created large cities and a strong centralized political structure; they developed specialized architecture, temples, pyramids, and many supportive arts. The early Sumerians of Mesopotamia were apparently the first to create a written language and to record the events of their life in a form that can be read today. With them, our knowledge of mankind passes from prehistory into classical history.

Cro-Magnon man became established over much of Europe some 30,000 years ago, long before the development of agriculture and village life. Although he lived chiefly as a hunter, he seems to have been very similar to us both anatomically and mentally; his tools and weapons fit our hands; his cave art moves our souls; the care with which he buried his dead reveals that he shared with us some form of ultimate concern. Every trace of prehistoric man in the world provides further evidence for the view that the fundamental characteristics of *Homo sapiens* have not changed since the Stone Age.

Despite the constraints imposed by the unity and permanency of man's biological nature, the manifestations of human life have displayed a rich diversity which was already apparent during the Paleolithic Age and naturally increased during the Neolithic Age, and especially during the Bronze Age, when human populations became more stable. There are still in Africa, Asia, Oceania, and South America a few small tribes whose ways of life hardly differ from those of Stone Age man. They eke a meager living out of the natural resources of their surroundings; some, like the Australian Bushmen, do not practice even the most primitive form of agriculture. But all these tribes have extremely complex languages, traditions, social structures, and religious beliefs, in addition to possessing an extensive practical knowledge of their environment. Clearly culture—if this word is defined as everything learned by experience and transmitted from one generation to the next—can reach high levels without elaborate technology.[5] Culture is the expression of man's responses to the physical and human environment. These responses take the form of behavioral patterns and emotional relationships as well as the development of utilitarian objects.

The interplay between man's nature and environmental forces is strikingly illustrated by the comparative histories of ancient Mesopotamia and Egypt.[6] These two lands lie close to each other in the Near East and have therefore much in common. Of special importance for their early economic and social development were the facts that each is centered about a great river valley and each blessed with a potentially fertile soil. In both areas, however, the agricultural land had to be created out of the wilderness by human toil and ingenuity. The annual floods provided both water and soil. But the land along the river was originally swamp and reedy jungle. To make it usable for crops was a stupendous task that required elaborate social organization. The swamps had to be drained by channels, the flood waters restrained by banks, the thickets cleared away, and the wild beasts brought under control. Furthermore, because of the peculiarities of the rainfall pattern in most of the Middle East, systematic irrigation was required before the agricultural potentialities of the land could be fully realized.

Both Mesopotamia and Egypt developed successful irrigation techniques and economic systems very early and thus were able to create great civilizations—the oldest in the annals of mankind. These two civilizations were in contact from the very beginning and advanced at about the same pace. However, despite the similarities of their origins and the extent of their contacts, Mesopotamian and Egyptian civilizations differed profoundly in spirit and in basic content. The prevailing orientation in Mesopotamia was cosmopolitan; in Egypt it was provincial. The Mesopotamians in their religious practices accorded much prominence to the sky, the sun, the moon, and the stars. The Egyptians, however, were more inclined to deify the animals in and along the Nile than the heavenly bodies. Students of the Middle East believe that these cultural differences can be accounted for in part by topographic considerations, in particular by the fact that the Nile Valley is narrow and confined on both sides by high cliffs, whereas Mesopotamia is much more open to the surrounding plains and consequently provided easier opportunities for communication with the outer world.

Prehistory and ancient history provide many other illustrations of the diversity and variability of human cultures. On the other hand, from what can be surmised of religious cults in Stone Age cultures, and from the discussions of political behavior by Plato or Aristotle, it is evident that the important characteristics of human societies have remained much the same for several thousand years. The manifestations of these characteristics have changed but societies continue to serve the same essential needs and aspirations. While civilization obviously conditions what man becomes, it does not significantly affect his biological nature; what changes is the social milieu. Habits and skills accumulate in society as in a reservoir and thus become available to human beings in successive generations. But, as the English historian Arnold Toynbee wrote: "Scratch the surface and efface what we receive from an education which never ceases and we shall discover something very like primitive humanity in the depths of our nature."[7] This is true not only for social behavior but also for biological and emotional needs. Cro-Magnon man, if he were born and educated among us, could work in an IBM plant and might even become president of the company. Modern man could readily return to primitive life, and indeed he does to some extent whenever he needs to.

The reader will observe that in the preceding paragraph I have used "civilization" in the singular and "cultures" in the plural. This deliberate distinction may help to convey the way I shall try—but frequently fail—to differentiate between these two words in the following pages. In principle, I shall use "civilization" when referring to the values that can be shared, and are increasingly shared, by most people irrespective of origin, race, or religion. In contrast, I shall use "cultures" to designate the body of values, ideas, and beliefs characteristic of a particular group. Science, the technologies derived from it, and certain ethical concepts are meaningful to most of mankind and might serve as the basis for a universal civilization. Cultural values differ from group to group, change with time, and imply the diversity inherent in mankind. As I shall repeatedly emphasize, universality and diversity are two complementary aspects of man's nature.

The Human Adventure in the New World

The prehistoric and historic events of the human adventure in the Americas vividly illustrate both the biologic unity of mankind and the experiential diversity of human life.

No evidence of prehuman anthropoid forms of life has been found anywhere in the New World, from the Arctic shores to Cape Horn. All Americans, including the Indians and the Eskimos, are descendants of immigrants. Just how long ago the first men arrived in the New World is still a matter of debate. Some of them had certainly become established on the continent 12,000 years ago; human artifacts recently discovered in Mexico appear to be much older, perhaps 40,-000 years old. In any case, one can surmise that man entered the North American continent from northern Asia over the Bering Strait during the last glaciation, and that there were several episodes of entry by independent groups. After the initial penetration, the immigrants rapidly spread over most of the two linked continents. Some of them had reached the extreme southern tip of South America at least around 7000 B.C., as shown by remnants of human occupation in two caves of the Fuegian area.[8]

The Stone Age Americans were primarily hunters, just like their Old World counterparts. Stone tools, and less frequently tools of bone, have been found in many areas, usually in association with the remains of now extinct animals, such as the mammoth and a large species of bison. As time passed, the hunters became most numerous in the Great Plains grasslands, perhaps because the large herds of bison and mammoth were restricted to this region when the forest returned in the wake of retreating ice.

Progressively, the various groups of immigrants followed several paths of further migration and settled in different parts of the continent. It can be postulated that they completely lost contact with Asia and other areas of their origin, and each group very soon also became separated from the other migrating groups spread over the Americas. Deserts, mountain masses, dense forests constituted numerous barriers between the Pacific and Atlantic oceans, causing the progressive emergence of a multiplicity of human groups that evolved independently and thus gave rise to different sub-races.

The populations of the New World eventually domesticated plants and animals, as had occurred during the

Neolithic period in the Old World. Maize, beans, squash, potatoes, and manioc were their main crops. Their animals were turkeys, ducks, llamas, guinea pigs, and dogs. Their agricultural resources were thus very different from those first available in southwestern and southeastern Asia. Neolithic agriculture in the New World, emerging independently, seems to have reached its full development, first in Mexico, Guatemala, and Peru, probably around 1500 B.C. These areas also produced the first great American civilizations, just as Old World civilizations first originated in the regions which pioneered Neolithic agriculture.

Even granted that a few ships from Polynesia and Europe did manage to reach the Pacific and Atlantic coasts in prehistoric times—a romantic but unproven possibility—the majority of the American aborigines remained completely isolated from the rest of mankind until the sixteenth century A.D. During that period of isolation they progressively developed several cultures that evolved almost independently one from the other, each being exquisitely adapted to the topography, climate, food resources, and other aspects of nature characteristic of a particular region in the Americas.

The arctic Eskimos hunted walrus and whale in kayak boats. The Indians of the northwest coast harpooned seal and walrus and fished salmon in dugout canoes. The California Indians fished, hunted, and collected shellfish and fruit on the shores and in the valleys. The southwestern Indians grew maize, squash, and beans near their pueblo settlements. The Great Plains Indians hunted the buffalo, fished in the tributaries of the Missouri, and also raised maize, squash, and beans. The Indians of the northeastern woodlands farmed and collected fruits and berries.

Profound cultural differences developed along with

these differences in hunting and agricultural patterns. The Incas in the Peruvian Andes, the Mayas in the tropical forests of Central America, the Toltecs and the Aztecs on the Mexican plateau created societies, ways of life, and religious beliefs that differed almost as much from one another as from those of the Eskimos, the pueblo people, or the plains people.

The influence of the environment on human life is particularly well illustrated by the unique pattern of culture that emerged without benefit of farming along the northwest coast of Alaska and British Columbia. In this region, the Indians were favored by natural resources such as salmon, wild fowl, game, and forest trees. This enabled them to reach high levels of social and artistic achievement without having to develop agriculture or any other form of productive economy. As far as is known, tobacco was the only crop that they cultivated at all systematically.

During the period corresponding to the early phases of the Christian era in Europe, the New World cultures created strong theocratic governments, sophisticated calendars and mathematics, marvelous architecture and other arts; the early Mexicans built a true urban complex which can still be seen today at the immense site of Teotihuacán near Mexico City.

All the early Eskimo and Indian cultures evolved independently in Africa, Asia, Europe, and Polynesia. Yet the achievements of the American aborigines are meaningful to all peoples of all cultures originating from outside the Americas. This is true not only of their tools but also of their temples, sculptures, pottery, basketwork, and even more remarkably of their love songs, legends of creation, and social structures.

The parallelism of the agricultural, social, and artistic

achievements in prehistoric America with those of Africa, Asia, and Europe constitutes perhaps the best evidence for the unity of man's nature. It proves that the most fundamental and universal characteristics of the human mind were fully developed by the time Paleolithic man first penetrated the American continent and did not undergo significant changes during the thousands of years of isolation that followed. What could demonstrate more clearly the biological and psychological unity of man than the fact that the Spanish conquistadors married Indian princesses shortly after their arrival in Mexico and Peru?

In South America, as well as in other parts of the world, there still exist today small tribes which have had no significant contact with modern civilization and whose ways of life have hardly changed since the Stone Age. Yet experience has repeatedly shown that individual members of such tribes, born and raised in a pre-technological environment under extremely primitive conditions, can nevertheless rapidly adapt to modern life, and acquire complex technological skills. An infant born to culturally backward parents but adopted very early in life by a more advanced cultural group or by one with a very different social tradition takes on the behavioral characteristics of the foster society and commonly rejects the culture of his natural parents when he comes in contact with it later in life.

Many Indians have of course refused to adopt the ways of Western civilization or have acquired only its worst aspects. But this is due to cultural and social reasons rather than to biological ones. Countless Indians from all parts of the Americas are now completely westernized and perform as well as the people of European origin in positions involving complex technological or administrative responsibilities. An appealing example of adaptability is described at length in the biography of Ishi, a California Indian who was the only survivor of a small isolated tribe still living in Stone Age culture, and who was adopted in his thirties by the Department of Anthropology at the University of California in San Francisco. Ishi soon learned to command an English vocabulary of some 600 words and could understand many more words than those he himself used. He was employed as a general handy man around the anthropological museum and came to look forward as much as a white man to his afternoon automobile drive![9]

During the past five centuries, many waves of immigration have brought a multiplicity of different types of human beings to the American continents. Europeans, Africans, and Asians have thus been brought into intimate contact with the various tribes of American aborigines. Many varied and highly successful racial mixtures have resulted from interbreeding. The genetic and physiological compatibility between races that had been separated for so many thousands of years confirms the cultural evidence that all human beings originally derive from the same evolutionary stock, have retained the same fundamental nature, are compatible biologically and mentally, and seem to be endowed with the same potentialities.

While man's nature exhibits such remarkable unity and permanency, his social institutions and ways of life are extremely diverse and changeable. Cultures differ with time, and from one place to another, because human beings are endowed with a wide range of potentialities which allow them great latitude in responding to environmental circumstances. We shall now illustrate briefly some environmental determinants of these responses.

In his essay on *Airs, Waters, and Places* the Greek physician Hippocrates boldly asserted 2,500 years ago that the physical and temperamental attributes of the various populations in Europe and Asia were determined by the physical characteristics of each particular region, such as climate, topography of the land, composition of the soil, quality of the water.[10] Health, endurance, physical stature, military prowess, and even political institutions were commonly regarded by Greek physicians and philosophers as direct expressions of man's responses to such environmental factors. A number of modern anthropologists and sociologists have enlarged on this Hippocratic concept by pointing out that, in the past at least, agricultural potentialities have played a large role in determining cultural development as well as limiting its manifestations.[11]

The failure of the Inca Empire to extend its boundaries into regions with lesser agricultural potential has been quoted by anthropologists as a typical illustration of the dependence of advanced cultures on a highly productive subsistence base. Populations living in areas where the soil, altitude, temperature, rainfall, or growing season are unfavorable to farming have rarely succeeded in developing cultures that went beyond the nomadic stage, with primitive social organization and limited material equipment. All exceptions can be traced to unusual circumstances, such as the bountiful and permanent supply of wild plants, fish, game, and wood available to the Indians of the northwest coast of North America, or the local conditions that permitted the development of an efficient pastoral technique of food production in certain parts of Asia.

Needless to say, physical surroundings, agricultural development, industrial resources, and other economic potentialities are not the only factors that play a role in shaping the characteristics of human beings and of their culture. The American Indians of the northwest coast can serve here again to illustrate that the theory of environmental determinism of culture should not be carried too far.

The seasonal aspects of the principal fish harvest along the northwest coast imposed periods of intense activity on the Indian populations and stimulated them to develop techniques for the preservation of foodstuffs. The storage of large food supplies in turn made possible lengthy periods of leisure which the Indians used for ceremonials and the making of objects of art and luxury. The specific manifestations of these activities, however, were determined by historical and social factors rather than by the physical and economic environment. Similarly, the fact that the forests of the northwest coast provided an abundance of readily worked woods offered an inducement for indulging in sculpture and woodcarving crafts, but the subjects illustrated by artists and craftsmen were culturally determined.[12]

Many Indians of the coast are now commercial fishermen and loggers, more at home with gasoline and Diesel engines than with their ancestral canoes. The ancient arts and crafts have all but disappeared, in part because the Indians no longer believe in their ancestral gods, but also because they do not find the time to carve and to paint now that they have accepted the efficient ways of technological civilization! Competitive societies, in which economic reward is considered the only measure of value, seem to destroy arts and crafts, whether they are those of medieval and Renaissance Europe or of the Indians of the northwest coast.

Religious and philosophical beliefs, social traditions, political institutions are among the many factors other than the physical and economic environment that determine the fate of human beings. Such factors act indirectly but powerfully. They govern the ways of life, affect physical appearance, are reflected in behavioral patterns, and through all these influences and many others impose a characteristic stamp on each particular culture. As we shall see later, all human characteristics are conditioned almost indelibly by early influences, namely the physical and social factors that impinge on the organism during the early, formative phases of development.

Man, furthermore, does not react passively to physical and social stimuli. Wherever he functions, by choice or by accident, he selects a particular niche, modifies it, develops ways to avoid what he does not want to perceive, and emphasizes that which he wishes to experience. The American Southwest provides a striking illustration of man's ability to create a way of life of his own choice, whatever the nature of the environment.

On their immense and thinly populated reservation, the Navajo Indians have long led a pastoral existence, taking shelter in isolated and temporary hogans, herding their sheep and goats on the semidesert land. In the same general region, the Hopi Indians live in adobe houses crowded in compact villages. They are land-poor, cultivate a few crops, and make the most of their scant water resources. The Navajos and Hopis are the products of the same land and the same climate but they have different traditions, worship different gods, and extract from their environment different ways of life. The American Southwest also provides a congenial environment for American ranchers who scorn the Indian traditions and for Mexicans who try to climb up the ladder of Western prosperity. Catholics, Protestants, Jews, and Mormons, as well as godless men of all races in search of adventure, find in the physical environment that has nourished the Navajos and Hopis other resources for creating cultures independent of the Indian past and indeed incompatible with the Indian ways of life.

At about the time in the mid-nineteenth century that settlers were creating the city of Seattle, English-speaking white men created the city of Victoria on Vancouver Island on the other side of Puget Sound. Vancouver Island became part of the British Colony of Canada, whereas the territory of Washington became part of the United States. Victoria and Seattle emerged as cities at the same time, they have approximately the same climate and the same natural resources, but their social and economic evolutions have been profoundly different. Victoria still cultivates a conservative and socially proper attitude reminiscent of nineteenth-century England. Seattle, in contrast, was dominated from the start by the enterprising spirit of the expanding frontier and acquired the dynamic attributes of the twentieth-century United States. The difference between the sedateness of the Victoria Britishers and the euphoric aggressive optimism of the Seattle Americans does not originate from differences in geographical conditions, in economic resources, or in racial characteristics. It is the expression of human choices which in turn can be traced to social history.

Human life is certainly influenced by environmental factors and it is also conditioned by the past. Even more interestingly, however, the life of a particular person, or of a social group, becomes to a very large extent what he or the group wants it to be, through a succession of deliberate choices. History, both individual and social, is the account of the ways by which men meet the challenges of their environment through the instrumentality of their innate endowment, steered at every step by the vision of their goals.

Progenitors and Contemporaries

Whether he is blond or brunet, tall or short, and whatever his social background or occupation, the typical Englishman differs in many behavioral characteristics from the typical German, Italian, or Frenchman. The differences extend from the manner of walk or the use of the hand to familial and social relationships; from eating and drinking habits to the manifestations of religious worship or of anticlericalism. Belonging to a certain nationality means much more than allegiance to its constitution and its flag. In an ill-defined but very real way, most nationals of a given country have in common many characteristics that transcend physical stature, geographical origin, and even race. Two illustrious leaders of the United States Armed Forces during the Second World War, General Eisenhower and Admiral Nimitz were descended from German immigrants, yet there was nothing to reveal a foreign origin in their physical appearance and behavior. Similarly, the general who led the Free French into Paris at the end of the war was called Koenig but despite his German name he was French to the core. Each human being is the product of the physical and cultural environment in which he has developed at least as much as of the genetic endowment he has inherited.

While individual persons are shaped by the environment in which they develop, history shows that in any given country certain national characteristics persist with only minor changes for many generations. This stability results in part from the relative constancy of the physical environment in a given area, but even more from the fact that the cultural environment is self-reproducing and thus maintains a continuity of social influences.

But this does not mean that the physical traits and behavioral patterns of a given national group are unchangeable. The swinging English teenagers and young adults of today differ physically and mentally from the pompous Tories and debilitated Cockneys of the Victorian era. Russians and Italians in the second half of the twentieth century exhibit a technological and economic dynamism that would have been difficult to predict from the moods and attitudes of their respective countries two generations ago. Men are conditioned by genetic and environmental factors, but they can escape from the bondage of this conditioning.

In his essay on "The Uses of Great Men," first published in 1876, Emerson expressed succinctly but powerfully the importance of environmental factors in shaping human characteristics. "There are vices and follies incident to whole populations and ages. Men resemble their contemporaries even more than their progenitors."[13]

Men resemble their progenitors because they derive from them not only their genetic endowment but also many of the attributes that are acquired in childhood by learning and experience. They resemble their contemporaries because, within a given country and social group, most members of a given generation are simultaneously and similarly influenced during their development by the prevailing environmental factors. When parental influences are dominant, as in conservative and stable societies, men tend to resemble their progenitors. But parental influences are likely to be overshadowed by other environmental forces when the ways of life are rapidly changing, as was the case in America when Emerson wrote his aphorism.

In practice, the human genetic pool remains essentially the same from one generation to the next. Any change in human characteristics must be traced therefore to the upbringing of children and to other influences exerted on them by the total environment. A few brief examples will suffice to illustrate how profoundly and rapidly changes can occur in the biological characteristics of a given population.

In the affluent countries that have adopted the ways of Western civilization, teenagers now achieve physical and sexual maturity much earlier than did their counterparts of earlier generations.[14] Similarly, many people now reach larger adult size than was the case in the past. In certain countries, Japan for example, such changes have taken place since the end of the Second World War, within one generation. The rapidity of these changes excludes a genetic mechanism and proves that they represent phenotypic manifestations of man's responses to the contemporary environment. Another aspect of the modern scene in which contemporary man appears different from his progenitors is the pattern of diseases. Tuberculosis and some types of neuroses that were most common in the Western world at the turn of the twentieth century have now all but disappeared, to be replaced by other disorders, such as vascular diseases, lung cancer, and drug addictions. The prevalence and severity of disease commonly change from one generation to the next. Each type of civilization has its own pattern of disease.[15]

Catastrophic events, such as droughts and harsh winters, famines, epidemics, and wars, loomed very large in the past history of the human race. In most parts of the world, especially in underdeveloped countries, the effects of such natural forces are as important as ever. More and more, however, man is now responsible for the introduction of new environmental factors that condition and often threaten all aspects of his life. During the early phases of the Industrial Revolution, infectious processes, nutritional deficiencies, hard physical labor, and the sudden migration from sparsely populated areas to congested and unsanitary urban environments were among the factors that affected the proletariat most severely. The bourgeois classes had their own problems, originating from overeating, lack of physical exercise, and other misuse of economic affluence, as well as from the psychological constraints created by unreasonable social conventions. In our own societies, the influences of material wealth and of the industrial environment are now compounded by the effects of generalized urbanization.

Barring nuclear warfare or some other global cataclysm, the world population will continue to increase for several decades at least, in affluent as well as in underdeveloped countries. This will happen even if contraceptive techniques achieve universal acceptance.[16] With the low mortality rates now prevalent in all countries that have introduced modern public health practices, the population can be stabilized only if the number of children is less than 2.5 per couple. There is no evidence that family control will soon reach this drastic level anywhere. It can therefore be taken for granted that the world population will greatly increase in the immediate future and will indeed probably double within less than a century. As a consequence, the largest percentage of human beings will be born and develop, and their children will be born and develop, within the confines of large urban agglomerations. Whatever individual tastes may be, mankind will thus be shaped by the urban environment.

Before considering the biological consequences of this fact, I must point out that the urban environment now includes and will probably be increasingly made up of suburban districts. In these areas, the ways of life are almost completely urbanized, even though population density is much lower than in the compact city. The move to suburbia has many motivations other than the desire for space, quiet, and greenery. In many cases, it constitutes an attempt to escape from some of the constraints of city life and to recapture the pastoral or village atmosphere of the traditional past. Suburbanites cultivate the illusion that the detached house with its garage and front lawn is almost the equivalent of the farm, with its outbuildings and pastures. Since the suburban area is divided into a multiplicity of districts, each with its own zoning board, schools, water and sewage systems, and fire department, all more or less independent of one another, the district management provides suburbanites with some chance to perpetuate the town-hall tradition of administrative autonomy. The suburban environment is thus identified in principle with a formula of social life somewhat different from that provided by the city apartment house. The difference, however, is rather superficial.

The suburbanite may have a lawn in front of his house, but the air he breathes, the water he drinks, and the food he eats are as chemically processed or polluted as those of the city apartment dweller. He may have a tool shop in his basement, but he is as completely dependent on public services as if he lived in the heart of the city. He may consider his home an inviolable castle, but he experiences crowding, traffic jams, aggressive competition, and social regimentation wherever he goes for work or for leisure. The very design and decoration of his living quarters are governed by the need to accommodate equipment that he does not understand; all his activities expose him to stimuli very different from those under which human evolution took place. He is increasingly dissociated from the cycles of nature that have established the biological rhythms of human life and that have shaped its physiological functions. In fact, most of his contacts with the outside world originate from technology or are mediated through technology. To adapt a phrase from the American literary critic Leo Marx, the machine is in his garden.[17] To discuss the effects of the city or suburban environment on human life is in practice tantamount to discussing the consequences likely to result from the transformation of the modern world by scientific technology.

Crowding, regimented life, environmental pollution, and disturbances of the fundamental biological rhythms are aspects of life which are common to all highly technicized and urbanized societies, rich or poor. These influences elicit from the human organism responses from which are emerging the physical, mental, and social disorders commonly called diseases of civilization. These responses impress a characteristic stamp on modern life. They account for the fact that Emerson noted—we resemble our contemporaries even more than our progenitors.

Let me emphasize again that the radical changes in growth, health, and behavior that result from life in the urbanized, technologically controlled environment are not caused by genetic disturbances. In practically all cases, the changes represent responses of the human organism to environmental stimuli.

Cultural evolution has long been of much greater importance than biological (genetic) evolution, but this does not mean that the biological evolution of mankind has completely stopped, or that it is irrelevant to cultural evolution. The very fact that cultural forces are changing the ways of life so profoundly and rapidly makes it certain that genetic effects on man himself will eventually follow.[18]

Knowledge of human genetics, imperfect as it is, leaves no doubt that there constantly exist in human populations a number of genetic variants, associated with different degrees of fitness for one or another type of environment. Genetic factors providing a higher fitness for new environmental circumstances naturally tend to be favored by selection and progressively become more prevalent. Since human populations have a wide range of potentialities, the selection of variants can probably change the genetic structure more rapidly than used to be thought. For example, continued life in crowded cities might increase within a few generations the resistance through genetic mechanisms to crowding, noise, and social regimentation.

Since the environment in which man functions and multiplies is primarily the sociocultural environment that he creates, the genetic evolutionary changes most likely to emerge out of a certain form of culture are the ones that increase man's fitness for this very culture. Moreover, the genetic changes that increase the fitness of their carriers for a particular culture also increase dependence on that culture; they stimulate thereby further cultural developments and these in turn instigate further genetic changes.

Evolutionary changes can favor socially undesirable genetic characteristics just as well as the ones that are desirable. Since the human beings best adapted to social regimentation are likely to have an advantage in a crowded world, their biological success may help mankind to survive the population avalanche, but this will accelerate the movement of our societies toward the conditions of the ant hill.

The sociocultural forces that condition both the contemporary expressions of man's genetic endowment and the future course of his genetic evolution have their origin in the distant past of the human species. Some of these forces can be tentatively traced back to the Stone Age when man acquired many biological needs and urges that he has retained until now; many social structures, mental attitudes, and ways of life that are ubiquitous in modern societies also have such distant and vague origins. Other sociocultural forces can be more precisely recognized as beginning in Mesopotamia at the very dawn of historical times. Some 10,000 years ago the Sumerians created in Mesopotamia the first great civilization and the first written language of which we have knowledge. The following quotation, taken from a scholarly analysis of the emergence of culture in the Near East, gives some idea of the richness of Sumer's legacy to us, and of the extent to which our present life is still influenced by Sumerian achievements.

"When we of today reckon our years by the sun, our weeks by the moon, and identify our days after the planets; when we look at our timepieces to tell the hours, the minutes, and the seconds, in conformance with a circle of 360 degrees and the sexagesimal system of counting; when we break up the natural sciences into their component disciplines; when we approach the babel of known languages through the medium of internal analysis; when we write our official documents, our scientific calculations and conclusions, our literary creations, and our private letters; when we reaffirm our belief in laws impersonally conceived in a government that is a safeguard against autocracy—when we do these and countless other things, we are utilizing, consciously or unconsciously, the results of an immemorial experiment in living in which ancient Mesopotamia played a leading part throughout the first half of recorded history....

"This significant attainment outlasted the parent civilization itself, as well as its numerous clients and successors, and it entered eventually, enriched by the independent contributions of Israel and Greece, into the mainstream of Western Civilization....

"The Hebrews and the Greeks are closer to us in modes of thinking. Yet the Sumerians and the Egyptians opened the door to civilized times and gave us forms and institutions which we still take for granted: family life, government, law, social behavior, writing, education, and the beginnings of science. They gave us chairs and tables, villages and houses, tools and weapons of metal, and a fully-structured architecture. They gave us a calendar and a formal art and literature of high complexity."[19]

The Mesopotamians also transmitted to us many of their social problems. Because of the crucial role that water played in the early history of the Near East, conflicts continuously arose over water rights. Etymologically the word "rivalry" derives from the Latin *rivus*, a stream. The history of Mesopotamia is replete with quarrels over competing canal systems. The recent struggles over the Suez Canal and the Gulf of Aqaba, not to mention those over the Dardanelles and the Panama Canal, have made clear that we are still often struggling over water rights and that in more ways than one rivalry can be traced back to Sumerian civilization.

Man's very ancient biological nature first took a sociocultural form among the Paleolithic hunters, then among the Neolithic farmers, and finally among the Sumerians in Mesopotamia. This biological and cultural heritage is indelibly incorporated in all subsequent activities and achievements of mankind, including our own. It has undergone endless transformations in the course of time and in different places, but has persisted as an indestructible component of all the civilizations it has generated.

The historical determinism of social life is not limited to the influences of the ancient past. Each developmental stage of any particular culture usually retains many components of the preceding stages. For example, it is easy to recognize historical influences in the comparative evolution of the urban environment in Europe and North America. The classical city of the European tradition was compact and its inhabitants were apartment dwellers. Such is still the case today for Rome, Florence, Paris, Hamburg, and other continental cities (although less so for English cities). In contrast, the typical American city is sprawling; its inhabitants live in individual detached houses with a lawn that often exceeds in size the land available to a Chinese farmer; its loose structure reflects the agrarian and nomadic tradition of American life. This is true even of huge cities—Greater New York and especially Philadelphia—since a very large percentage of their inhabitants live in detached one-family houses that they own.

The contrasting traditions of urban development in Europe and America are symbolized by the personalities of two great contemporary masters of architecture and planning. The French-Swiss architect Le Corbusier was influenced by the European classical tradition when he visualized his Cité Radieuse, consisting of immense self-contained dwellings located in the midst of disciplined parks. The American

Frank Lloyd Wright in contrast expressed the pastoral tradition of America when he advocated the Broadacre type of planning with one single-family house per acre. Precisely because antecedent social forces still condition the development of all aspects of civilization, each part of the world tends to retain its cultural identity, despite the fact that raw materials, technological practices, and power equipment exhibit such uniformity all over the world.

Except under unusual circumstances, man tends to accept the traditions of his group as embodying the truth. Even when he rebels against them, the attitudes he takes and the new ways he tries always incorporate many of the ancient traditions and thus keep him dependent on his social and cultural past. All revolutions are national in character, especially when they label themselves international. In each country, likewise, the rebellious youths and all the forms of lost, beat, and "now" generations display patterns of behavior that reflect some aspects of their cultural origin. This probably explains why most rebellions fail, since rebels find it so easy to return to the fold.

Knowledge of the past is essential for the understanding of life in the present and in the future, not because history repeats itself—which it never does exactly—but because the past is incorporated in all manifestations of the present and will thereby condition the future. At every stage, human life is the incarnation of the past. The Spanish philosopher José Ortega y Gasset (1883–1955) expressed, in his famous aphorism, "Man has no nature. What he has is . . . history," his conviction that the past shapes the behavior of human beings.[20]

Man is indeed the product of his history, but history is

far more complex than Ortega's statement would suggest because it also includes biological determinants. The bodies and the minds of individuals and the expressions of social life in the various cultures are the living records of biological influences that have been constantly at work from the most distant past until the present time. Some of these influences have left their stamp on the genetic make-up of each individual person, others on the physical and mental characteristics he acquires during life, still others on his social structures. Humanity continues to grow by incarnating the past.

3.

BIOLOGICAL REMEMBRANCE OF THINGS PAST

The Genetic Record of Past Experiences

In its most general sense, the word "evolution" means the progressive transformations of a system in the course of time. When biologists or sociologists use the word, they usually have in mind the long-range molding of living organisms or institutions by the environment. Biological species, individual persons, societies or their institutions are indeed molded by the environment as a result of the adaptive responses that they make to its stimuli. Considered broadly, evolution always involves learning from experience. The learning may take place by storage of genetic information in the chromosomes, by accumulation of knowledge and skills in the individual organism, or by transmission of practices and wisdom in institutions or in society as a whole.

Neo-Darwinian biologists give to the word evolution a more precise but narrower meaning. For them, it denotes the transformations of a species resulting from spontaneous mutations in its genetic equipment and from the selection of mutants by environmental forces. Darwinians regard natural selection as the agency that translates environmental challenges into the genetic alterations of the species and thus brings about the evolutionary changes which improve its fitness to the environment. Since the following discussion is focused on Darwinian evolution, it may be helpful to define briefly some of the determinants of the evolutionary process.

The amazing ability of living organisms to learn from experience and to transmit this learning to their progeny is greatly facilitated by certain peculiar characteristics of their genetic equipment. On the whole, genes are very stable structures and thus they usually transfer the hereditary attributes of the organism from one generation to the next in an unaltered form. But equally important is the fact that genes are not completely stable. As they spontaneously undergo alterations now and then, the species can respond adaptively to environmental changes by using the mutant forms thus produced. Sexual reproduction also facilitates the evolutionary process. Sexual union results in a shuffling of the two different gene arrays contributed by the two mating organisms and thereby brings about the emergence of new characteristics through new combinations of genes. The availability of these combinations in turn enlarges the range of adaptability of the species to changing environmental conditions.

Heredity does not determine fixed characters or traits; it only controls developmental processes. Furthermore, the path followed by any developmental process can in principle be modified by both genetic and environmental variables. However, the degree of modifiability or plasticity is quite different for different processes. As a general rule, the processes essential for survival and reproduction are buffered against environmental and genetic disturbances; in other words, they are not readily affected by the environment. Two eyes, a four-chambered heart, the ability to maintain an approximately stable body temperature, the suckling instinct in the infant, sexual drive in the adult, the capacity to think symbolically and to learn a symbolic language are all characters that develop in almost every human being irrespective of the environment in which he lives. Their development is coded in the genetic constitution in such a manner as to be little affected by external factors. In contrast, less stable characters are generally those for which variability is advantageous. For example, sun tanning and shade bleaching are obviously brought about by environmental factors.[1]

Under natural circumstances, mutations are spontaneous events, shuffling of genes during sexual reproduction is accidental, and the selection of new genetic assortment by environmental forces occurs blindly. Genetic evolution is therefore an unconscious process. The interplay between organism and environment, however, is far more subtle than is indicated by this simple formulation, particularly in the case of the higher animals and man.

It is a truism that a given environment can act as a selective agent, and thus govern evolutionary changes, only if the animal elects to stay in it long enough to reproduce. In general, an animal occupies a given site and continues to function there because forced to do so by external forces. Commonly also, the animal reaches a new environment accidentally in the course of exploration and elects to remain in it. Such choice implies a preadaptation which can be either genetically determined or the result of prior individual experience. In any case an important aspect of the preadaptive state is that the natural selection exerted by the environment is preceded by some kind of choice, not necessarily conscious, by the animal. Whatever the precise mechanisms of this ill-defined situation, common sense indicates that animals and men do not behave as passive objects when they become established in a given environment.

Eventually a change of environment leads to a change in habits, which in turn modifies certain characteristics of the organism. Even when repeated for several generations, such modifications are not truly inheritable, but they may nevertheless foster evolutionary changes. The reason is that continued residence in a particular environment tends to favor the selection of mutants adapted to it. Eventually, such mutations are incorporated into the genetic structure of the species involved.[2]

The preceding inadequate statement of an immensely complex problem will suffice to indicate that the evolutionary system comprises not only the mutations, genetic recombinations, and selective processes of classical neo-Darwinism, but also the processes by which animals, and especially men, choose and modify one particular habitat out of all the environmental possibilities available to them. This view of biological evolution does not imply acceptance of the discredited Lamarckian hypothesis, according to which changes in physical or mental activity, if continued long enough, eventually bring about corresponding changes in the body or the mind that are genetically transmissible from parent to offspring. Biological evolution always takes place through the spontaneous production of genetic mutants which are then selected by environmental forces. But the actual selective mechanisms are extremely complicated; they always involve uninterrupted feedback processes between the organism, its environment, and its ways of life.

The exquisite adjustment between the shape of flowers and of the birds or insects that pollinate them by feeding on them beautifully illustrates the place of the feedback concept in the theory of evolution. From the comparative study of fossils, it has become apparent that primitive forms of the flowers evolved simultaneously with primitive forms of their bird or insect visitors. Mutual anatomical adaptations developed progressively through countless small adaptive changes occurring over millions of years of continuous interplay.[8]

The evolutionary development of most animals has certainly been profoundly influenced by their ability to move, to learn and to establish social structures. While this is difficult to demonstrate convincingly, observations on birds and mammals in the wild and in the laboratory point to a variety of mechanisms that may have a bearing on human evolution.

Certain evolutionary changes probably had their primary origin in an exploratory curiosity that made animals discover new ways of sustenance and of life. In Great Britain during the past few years, the birds known as tits have developed the habit of pecking through the cardboard tops of milk bottles delivered in the morning at doorsteps. Apparently, the birds open the bottles to get at the cream. As one tit tends to imitate another, the habit has progressively spread from a few centers in Britain to other parts of Europe. It does not seem unreasonable to postulate that individual members of this species endowed with a gene complex preadapted to bottle opening would have a better chance of survival in urban environments, and thus would initiate an evolutionary trend. Through selection of mutants, a new shape of beak might evolve as a result of the change of habit, provided milk containers and the practices of milk distribution remained the same long enough to continue favoring the birds which have become specialists in bottle opening.[4]

Field studies and unpublished experiments by my colleague Peter Marler provide another intriguing suggestion for the possible role of learning in the evolution of birds.

Under natural conditions, birds learn their song patterns from their parents and from other birds of the same species around them. In the laboratory, newly hatched birds can also learn from playbacks of recorded songs. When male white-crowned sparrows are raised in isolation and exposed to sequences of two recorded songs, one of their own species and one of another species, they invariably learn their species song. This selectivity explains why most birds in the wild learn only one kind of song and why the songs of all male birds of one species living in the same area are almost identical. The selectivity, however, is not absolute. Striking differences in vocalization patterns occur consistently among birds of the same species from one area to another, even within small distances. These differences are so stable from year to year that ornithologists are wont to speak of bird song "dialects."

Young white-crowned sparrows can learn a dialect other than that of their parents if they are raised in isolation and exposed early in life to the song of adult birds of their species obtained from another area or to recorded playbacks of such songs. In the wild, the variations in vocalizing patterns must be transmitted from generation to generation, young birds learning from their parents.

No anatomical differences have been recognized among birds obtained from different dialect areas. The stability of the dialects in nature suggests that, despite the mobility of birds, little exchange of individuals between populations occurs after the vocalizing patterns have been acquired. If males are more likely to settle in an area where their dialect is heard, and if females are more likely to mate with a male living in their own dialect area, the population will be progressively fragmented into small local subgroups and inbreeding will occur. Inbreeding continued over long periods of time almost inevitably brings about hereditary changes. Thus, as small a behavioral difference as bird song dialects, first entirely under environmental control, may eventually be incorporated in the genetic endowment of the isolated populations.[5]

Populations of primates have been repeatedly observed in nature to learn entirely new habits from one of their members—for example, washing food or unwrapping and eating caramels.[6] Changes in habit may eventually alter the structure of animal societies and affect genetic constitution through selective processes. Because the diet of chimpanzees consists largely of large fruit, in the wild they must move over wide areas to find enough food and therefore cannot live in stable social groupings. In contrast, gorillas eat almost any plant food at hand in the rich and varied vegetation of their tropical environment; since they need not move far from their home base they can form permanent family groups.[7] It has been suggested that the differences in food habits and social structure between these two primate species have influenced their genetic evolution.

The environments into which man's precursors moved during the Paleolithic periods naturally conditioned the activities in which they engaged, and this in turn must have contributed to determining the genetic endowment which defines mankind.[8] Similarly, the habits introduced by modern civilization are now acting as selective forces and guiding human evolution toward new forms of adaptation to the technicized urban environment.

Under certain conditions, in contrast, cultural attitudes can oppose the selective effects of the physical environment. In the Indian city of Cochin, for example, the famous street of the White Jews harbors a Jewish population that has lived there for nearly 2,000 years. In contrast to the rest of the Cochin population, these Jews have retained skins as white as those of their ancestors who migrated from Palestine, the reason being that they have carefully kept themselves out of the sun's direct rays under roofs and lattices. At the present time, only 150 Jews out of a former population of 1,500 still exist in Cochin, but despite their small numbers, they have retained their biological identity through environmental protection, choice of occupation, and a breeding pattern of their own choosing enforced by religious sanctions.[9] They illustrate the fact that culture is as much a part of the total environment as solar radiation, temperature, rainfall, or altitude.

All living organisms retain structural and functional evidences of their distant evolutionary past. Whatever the conditions under which they are born and develop, their responses to stimuli are always affected by the experiences of the past which are incorporated in their genetic make-up. The evolutionary steps through which man reached the level of *Homo sapiens* explain, for example, why the structure of his backbone can be traced to the early fishes, or why the salinity of his blood still reflects the composition of sea water from which terrestrial life originally emerged.

The thickening of the sole of the foot relative to the rest of the skin probably constitutes an expression of biological remembrance of the past, since this process starts before the body experiences any frictional stimulus and is detectable even within the womb. It seems legitimate to assume that, as the protoamphibian ancestors of man occasionally came out of water and pushed themselves on land with their fin lobes, these organs responded by a minor thickening on the part of the skin that came in contact with the ground, much as our skin thickens whenever it is subjected to friction. When locomotion on land became more common and the ability to develop calluses on the feet thereby became important for biological success, the individuals best endowed with this ability probably had a better chance of survival. The tendency to develop callus thus became inscribed in the genetic code.[10]

Many other types of anatomical and physiological attributes now considered characteristic of certain human races also demonstrate the persistence of the biological past. In the highlands of East Africa, where the fossils of early man are found in greatest profusion and where *Homo sapiens* probably emerged as a distinct species, the climate is moderate and provides a physical environment very similar to that which most human beings generally consider desirable for health, comfort, and activity. Granted this general preference for a temperate climate, certain adaptations have naturally occurred among the various sub-races of men under the influence of the local conditions prevailing in the regions where they settled over long periods of time. The fact that men who hunted on the grasslands and deserts must have been rigorously selected for speed may explain why a light, lean body build is still prevalent in the desert countries. In contrast, speed over long distances is of little use in a forest. Whereas the Arabs of the desert are likely to be tall and lean, typical forest dwellers are short-legged, long-trunked, barrel-chested, and broad-handed.

Subcutaneous fat is most abundant in people who have evolved in cold regions; differences are also found in the number of sweat glands per unit of skin area. Pigmentation of the skin is a function of exposure to light. While skin pigments are of particular importance for protection against ultraviolet light, it is also possible that the capacity of black skin to absorb visible light rather than reflect it, and to convert this light into heat, may also lower the heating threshold of dark-skinned races.[11]

In many animal species, the chemical changes in the sex glands that occur as a response to the environmental changes associated with spring initiate the process of courting and display. In birds, for example, this process is followed by nest-building, which begins at the proper time with the choice of the right material. Mating and egg-laying follow, then breeding and the feeding of the young. All these behavioral patterns are under the control of hormonal secretions which must be closely integrated with seasonal changes in the environment if the biological functions are to be successful. Such integration—the outcome of long evolutionary adaptation—has been most extensively studied in birds and in a few other animal species. In human life also, many physiological processes are still linked to cosmic events.

Modern man is wont to boast that he can control his external environment and has thereby become independent of it. He can illuminate his rooms at night, heat them during the winter, and cool them during the summer; he can secure an ample and varied supply of food throughout the year; he can if he wishes to make each day like every other day. But even when he elects to follow unchangeable ways of life in an environment which appears uniform, all the functions of his body continue to fluctuate according to certain rhythms linked to the movements of the earth and of the moon with respect to each other and to the sun. His hormonal activities in particular exhibit marked diurnal and seasonal rhythms, and probably other rhythms also linked to those of the cosmos.[12]

Man's physiological and behavioral responses to any situation are different in the morning from what they are at night, and different in the spring from what they are in the autumn. The writers of Western stories have a sound biological basis when they recount that the Indians always attacked at dawn, because they knew that the spirits of the white men were then at a low ebb. Napoleon is reported to have said that few are the soldiers who are brave at three o'clock in the morning. The wild imaginings of the night and the fears they engender are indirectly the effects of earth movements, because the human organism readily escapes from the control of reason under the influence of the physiological changes associated with darkness.

The lunar cycles are also reflected in the physiology and behavior of animals and probably therefore of man. It would not be surprising if moon worshipers—as well as "lunatics," as the term suggests—were really affected by lunar forces to which all of us are somewhat sensitive.[13]

Seasonal changes certainly affect most living things, including man, even when the temperature and illumination are artificially maintained at a constant level. In the most mechanized, treeless, birdless, and air-conditioned city, just as in the hills of Arcadia long ago, men and women perceive in their senses and reveal by their behavior that the exuberance of springtime and the despondency of late fall have origins more subtle than the mere change in temperature. It is for good biological reasons that Carnival and Mardi Gras are celebrated when the sap starts running up the trees, and that Europeans commemorate their dead in late fall when nature is dying. Modern man in his sheltered environment continues to be under the influence of cosmic forces even as he was when he lived naked in direct contact with nature.

The view that all aspects of life have historical determinants applies also to patterns of behavior that cannot yet be traced to hormonal activities or other physiological processes. Even the simplest organism differs from inanimate matter by virtue of the fact that all its activities are conditioned by its past. For example, the sea urchin displays a manifestation of biological remembrance when it responds to a sudden shadow falling upon its body by pointing its spines in the direction from which the shadow originates. Such response has a defensive value which is potentially useful because it helps in protecting the animal from enemies that might have cast the shadow. But in reality it refers to a past experience symbolized by the shadow—the possible approach of a predator—rather than the actual presence of a predator. The sea urchin's response to a shadow illustrates that even in the case of relatively primitive animals much of behavior is conditioned by ancient experiences of the species that have generated instinctive reaction patterns.

In the usual events of daily life, man continues to

react physiologically to the presence of strange living things, and especially of human competitors, as if he were in danger of being physically attacked by them. The fight-or-flight response, with all its biochemical, hormonal, and other physiological accompaniments is a carryover from the time when the survival of primitive man encountering a wild animal or a human stranger depended upon his ability to mobilize the body mechanisms that enabled him either to engage effectively in physical struggle or to flee as rapidly as possible.

On the other hand, man evolved as a social animal, and he can neither develop normally nor long function successfully except in association with other human beings. All social stimuli that man experiences elicit physiological and mental processes which in their turn condition his responses to the situations that evoke them. Thus crowding, isolation, challenge of any sort have effects that have their origin in the evolutionary past and that tend to imitate the kind of response then favorable for biological success, even when such a response is no longer suitable to the conditions of the modern world.

Many aspects of human behavior which appear incomprehensible or even irrational become meaningful when interpreted as survivals of attributes that were useful when they first appeared during evolutionary development and that have persisted because the biological evolution of man was almost completed about 100,000 years ago. Phenomena ranging all the way from the aberrations of mob psychology to the useless disturbances of metabolism and circulation that occur during verbal conflicts at the office or at a cocktail party are as much the biological expressions of the distant past as they are direct consequences of the stimuli of which the person is aware.

The urge to control property and to dominate one's peers is also an ancient biological trait that exists in the different forms of territoriality and dominance among most if not all animal societies. The lust for political power independent of any desire for financial or other material rewards, which is so common among men, has likewise prototypes in animal behavior. Even the play instinct and certain kinds of aesthetic expression correspond to derivative but nevertheless important biological needs that exist in one form or another among animal species and that have always been part of man's nature.

Jack London's story *The Call of the Wild* (1903) probably owed part of its popular success to its evocation of ancient precivilized traits that persist in man's nature. The actual story tells of a dog, Buck, taken from civilized life in California to be used as a sled dog in the Klondike. Jack London clearly intended to express through the dog's behavior his own protest against the constraints of civilized life. Some of the most beguiling moments of the story dramatize the power of the racial past, as Buck instinctively follows the lure of a wolf howl and finds a satisfying companionship among wolves far from men or dogs. The story ends with Buck plunging joyfully into the primeval forest to join the wolf pack, "leaping gigantic above his fellows, his great throat a bellow as he sings a song of the younger world."[14] Many psychologists interpret this story as advocating psychological regression, but it can be read with equal justice as a song celebrating the mysterious and wonderful world of the past which survives in the deepest layers of man's nature. It is dangerous to yield without thought to the call of the wild, but perhaps destructive to ignore it altogether. While the voices of the deep may seem strange and at times frightening, they are the expressions of forces that must be reckoned with, because inherent in the human race and influential in all aspects of human behavior.

Biological Freudianism

John Milton (1608–1674) was giving expression to common experience when he wrote in *Paradise Regained*:

> The childhood shews the man
> As morning shews the day.

The lines are valid as a general statement of fact, but they are also compatible with two entirely different theories of human development. One is that the characteristics of the adult person are the expressions of his heredity and are already apparent during his childhood; they merely continue to unfold during the rest of his life span. The other is that the experiences of very early life shape the physical and mental attributes of the child in an almost irreversible manner and thereby determine what the adult will become. The first theory is generally identified with the word "nature" and the second with the word "nurture."

There is no real conflict between these two interpretations. Both are correct, because each corresponds to one of the two complementary aspects of development in all living things. Whether the organism be microbe, plant, animal, or man, all its characteristics have a genetic basis, and all are influenced by the environment. Genes do not inexorably determine traits; they constitute potentialities that become reality only under the shaping influence of stimuli from the environment.

The present discussion emphasizes that the governing influence of environmental factors is particularly effective during the formative stages of life, both in the course of gestation and for a few years after birth. We shall see, furthermore, that effects of such early influences commonly last so long that they determine to a large extent the characteristics of the adult.

Long before biologists began to study the biological and mental effects of early influences, poets and novelists had derived some of their most universal and poignant themes from the depth and lasting quality of childhood experiences. Marcel Proust (1871–1922) struck a responsive chord in people of all races when he described how childhood atmospheres and events that had been forgotten could be brought back to consciousness in their original intensity by the trivial act of dipping a madeleine into a cup of tea. His *Remembrance of Things Past* is the literary expression of the biological truth that the memory of early experiences can be masked, but cannot be erased.

The experiments of the Russian biologist I. P. Pavlov in conditioning dogs to salivate at the ringing of a bell,[15] as well as all the subconscious Freudian complexes, constitute other well-documented illustrations of the extent to which responses during adult life are affected by childhood experiences. The Pavlovian type of conditioning can also occur during adulthood, but many forms of conditioning can be achieved only during certain "critical" periods of early life. To a very large extent, the practices of animal training, as for example with dogs, are based on the high receptivity to conditioning during these critical periods.[16]

The phrase "early influences" is most commonly used with reference to conditioning of behavior. Early experiences, however, do much more than condition behavioral patterns. They also affect profoundly and lastingly all physical and physiological characteristics of the organism at all stages of life. Experimentation with many types of animals and observations of human life have revealed that early influences affect not only learning ability, behavioral patterns, and emotional responses, but also initial rate of growth, ultimate size of the adult, nutritional requirements, metabolic activities, and resistance to various types of stress. In brief, most physiological as well as most mental attributes are lastingly affected by early influences.[17]

Many, if not all, effects of early influences are indeed so lasting that they appear to be irreversible. Few are the persons who do not recapture forgotten memories of atmospheres and events when exposed to fragrances or sounds first experienced during early youth. There is no dependable technique for erasing completely certain effects of early influences. It is possible, of course, that such techniques will be discovered, but it is also probable that some at least of the effects of early influences will be found to be truly irreversible. The more completely a biological system is organized, the more difficult it becomes to reorganize it. This is true whether the system is an embryo undergoing differentiation, or of a complex behavioral pattern that is becoming integrated. Prior organization inhibits reorganization.

As already mentioned, experiments in animals have shown that the young organism is particularly susceptible to the effects of conditioning during certain so-called critical periods of early development.[18] It is unfortunate that scientific knowledge concerning these critical periods is extremely scanty, because the same biological law certainly applies to human beings. Most slum children, unfortunately, continue to conform to the ways of life of their destitute parents, despite intensive efforts by skilled social workers to change their habits and tastes. By the third or fourth year of life their behavioral patterns have already been environmentally and culturally determined. Furthermore there is much reason to fear that they will in turn imprint similar patterns on their own children. It is not accurate to state that slum children are culturally deprived; the more painful truth is that slum life imprints on them a culture from which they are usually unable to escape.

The wide range of effects exerted by the conditions to which the organism is exposed during the early phase of its development can be illustrated by observations made on wild and domesticated animals. A spectacular experiment was carried out a few years ago by crossing the Shetland pony and the much larger Shire cart horse.[19] The birth weights of foals born to Shetland dams of Shire sires were similar to those of pure-bred Shetlands. Conversely, foals from Shire dams by Shetland sires had the same weight as foals of the pure Shire breed. The extent of intrauterine growth was determined by some form of maternal regulation. A factor of importance in this situation may be that, since the placenta of the Shire mare is three times as large as that of the Shetland mare, it provides a more abundant fetal nutrition, and thereby imprints the animal with a larger appetite after birth. There is evidence that, in the human species as well as in animals, the degree to which the maternal regulator constrains intrauterine growth reflects the degree of restriction experienced by the mother when she herself was a fetus. The process of growth control can thus spread over several generations.

It has been repeatedly observed that deer born during a period of food scarcity or in a crowded population remain of relatively small size even if their adult diet is adequate and steps are taken to decrease the density of the population. Again the effect of the nutritional deprivation can apparently extend to the following generation.

The lasting effects of early influences on the physical characteristics of animals have been demonstrated in many other types of experiments. The following are but a few among the many factors that have been manipulated in such experiments: nutrition, infection, temperature, humidity, type of caging, crowding or isolation, variety and intensity of stimuli. Whether acting indirectly through the mother during the period of gestation, or directly on the young shortly after birth, all these environmental factors have been found to affect the development of the young in a manner that is reflected in the adult.

One of the surprising facts revealed by such experiments is that almost any kind of stimulation of the young shortly after birth will affect the subsequent rate of growth as well as the resistance to stress. Daily handling, mild electric shocks, exposure to cold, and many other types of stimuli can thus have favorable or unfavorable effects. Activation of the hormonal systems of the young as well as changes in behavior of the mother toward her offspring are among the several mechanisms that have been invoked to account for the effects on growth and resistance.[20]

Anthropologists have found some indication that similar phenomena occur among human beings. They have observed that the stature of males seems to be increased by more than 2 inches in societies which make it a practice to mold or stretch the heads or limbs of infants. Piercing the ears, noses, or lips, cutting or burning tribal marks on the skin are other practices claimed to result in stimulation of growth. In fact, almost any manipulation or mild mutilation of the young, it is stated, may accelerate development and increase adult size.[21]

The most convincing knowledge concerning the effects that early influences exert on the characteristics of the adult has been obtained with regard to the nutrition of the mother during gestation and lactation, and of the young during early life. While all nutritional factors are certainly important, special emphasis has been placed on protein nutrition. Numerous experiments have shown that the rate of growth and the final adult size can be markedly depressed by limiting protein intake during gestation or the early phases of postnatal development. Similar results have been obtained by providing the mother or the young after birth with diets containing only proteins of low nutritional quality.[22] The depressing effect of early deprivation of proteins is aggravated by the fact that dietary habits acquired early in life tend to persist. Rats accustomed to a low-protein diet tend to continue eating it by preference even if a more nourishing diet is simultaneously presented to them later.

There exist all over the world countless situations in which human mothers and their babies experience inadequate protein intake and other nutritional deficiencies. Retardation of growth and small adult size are the usual consequences of almost any form of malnutrition or undernutrition during the early phase of development. After a time, moreover, the underfed human baby comes to accept without complaint nutritional intakes that are inadequate for optimum growth. The reverse of this situation may also be true. Some babies who have always been pressed to eat more than is necessary appear to become habituated to high food intake and to retain this need for the rest of their lives.[23]

The experiences of early life are particularly important in man because the human body and especially the human brain are incompletely differentiated at the time of birth and develop as the infant responds to environmental stimuli.[24] Japanese teenagers are now much taller than their parents and differ in behavior from prewar teenagers because the conditions of life in postwar Japan are different from those of the past. This finding is in agreement with the fact that first-generation Nisei children in America approach average American children in their growth and development. The children in the Israeli *kibbutz,* who are given a diet and sanitary conditions as favorable as can be devised, tower over their parents, who originated in crowded and unsanitary ghettos in Central and Eastern Europe.[25]

The acceleration of growth in Japan and in Israel constitutes but a particular case of the constant trend toward earlier development that has been evident in most westernized countries for several decades. This acceleration is evidenced by greater weights and heights of children at each year of life, and by the earlier age of the first menstrual period. In Norway the mean age of the onset of menstruation has fallen from 17 in 1850 to 13 in 1960. Similar findings have been reported from all affluent countries, and there is evidence that the advance in sexual maturation first took place in the well-to-do classes.

Not only is growth being accelerated; final adult

heights and weights are greater as well as being attained earlier. Some fifty years ago, maximum stature was not usually reached until the age of 29. In the affluent classes it is now commonly reached at about 19 in boys and 17 in girls.

The factors responsible for these dramatic changes in the rate of physical and sexual maturation are not completely understood. Improvements in nutrition and better control of childhood infections have certainly played a large part in accelerating development but other factors may also have been influential, one of these being greater facility of communication. The advent first of the bicycle, then of the automobile, decreased the tendency, almost universal in the past, for marriages to be contracted between members of a small community. Easier communication makes for larger number of acquaintances and consequently greater variety in the choice of marriage partners; the increased outbreeding results in what is technically called hybrid vigor. Still another possibility to be considered is the change of attitude toward children; for example, it has been established that the growth rate can be accelerated by such psychological factors as loving kindness at mealtime.[26]

If the trend in sexual maturation that has existed over the past 100 years had prevailed prior to that time, the age of first menstruation in medieval times would have occurred after the age of 30—an obviously nonsensical conclusion! In fact, there is suggestive historical evidence that the age of puberty in imperial Rome was about 13, at least among the affluent classes. The French biographer Pierre de Bourdeilles Abbé Brantome, writing in the sixteenth century, placed puberty at 12 to 13 years of age. The following quotation from Shakespeare's *Romeo and Juliet* also suggests that puberty was then reached almost as early as it is today.

Capulet: My child is yet a stranger in the world
She hath not seen the change of fourteen years;
Let two more summers wither in their pride
Ere we may think her ripe to be a bride.
Paris: Younger than she are happy mothers made.
(ACT I, SC. ii)

The advance in the age of puberty during our times seems therefore to be a restoration of the developmental timing which prevailed in the past and had been greatly retarded at the beginning of the nineteenth century. For lack of another theory, this retardation may be assumed to have been the result of changes in the ways of life and especially in the upbringing of children during the period of the Industrial Revolution. The complexity of the problem and our ignorance of its determinants are illustrated by the fact that the age of puberty in Britain now seems to be much the same in all economic groups, whereas height and weight are still linked to social class. In other words, acceleration of physical growth does not appear exactly to parallel acceleration in sexual maturity.

Experimental studies in animals have revealed that severe nutritional deprivations or imbalances during the prenatal or early postnatal period will interfere with the normal development of the brain and of learning ability. This has been proved by measuring the chemical composition and enzymatic activities of the brain as well as by a variety of learning tests.[27]

In man also, malnutrition occurring at a critical time appears to handicap mental development almost irreversibly. Of particular importance in his regard is the fact that the most crucial stages of brain development occur before and shortly after birth.

The growth rate of the human brain reaches a peak at about the fifth month of fetal life, is maintained at this maximum level until birth, and levels off at about the age of 1 year. At that time, the brain has achieved approximately 70 percent of its adult weight. By age 6 it is three times larger than it was at birth and its final structure is essentially completed. Language, thought, imagination, and the sense of self-identity have then reached a very high level of development. It is legitimate to assume, therefore, that the very structure of the brain and the fundamental patterns of its functions can be profoundly influenced both by the conditions of intrauterine existence and by the experiences of early extrauterine life.[28] Malnutrition during these critical periods almost certainly interferes with neuron development and as a result causes mental backwardness which cannot be corrected later in life. Certain infections of early life—including infantile diarrhea—may also affect irreversibly the development of the brain.

Many human beings make some kind of adjustment to malnutrition, but even when the adjustment appears satisfactory the remote and indirect consequences can be harmful. Recent physiological and behavioral studies have revealed that people born and raised in an environment where the food supply is inadequate in quantity or quality commonly achieve physiological and behavioral adaptation to low food intake. They reduce their nutritional needs by restricting their physical and mental activity; in other words, they become adjusted to malnutrition by living less intensely. Furthermore, they retain throughout their whole life span the physiological and mental effects of inadequate nutrition early in life.

Physical and mental apathy and other manifestations of indolence have long been assumed to have racial or climatic origin. But in reality these behavioral traits are often a form of physiological adjustment to malnutrition. Such adjustment has obvious merits for survival under conditions of scarcity; indolence may even have some romantic appeal for the harried and tense observer from a competitive society. But populations deprived during early life commonly exhibit little resistance to stress. They escape disease only as long as little effort is required of them and find it difficult to initiate and prosecute the long-range programs that could improve their economic status. They are prisoners of their nutritional past.

Malnutrition can take many different forms, including perhaps excessive artificial feeding of infants. Little is known of the physical and mental effects of a nutritional regimen which differs in quality from that of the mother's milk and exceeds it in the quantity of certain nutrients. Infants fed an abundant diet tend to become large eaters as adults. Such acquired dietary habits may in the long run have physiological drawbacks and it would be surprising if they did not have behavioral manifestations. Rapid growth and large size may not be unmixed blessings.

As far as can be judged at present, early development does not mean a shorter adult life; in fact, menopause seems to be delayed when puberty is advanced. Whether the acceleration in physiological development increases behavioral difficulties among teenagers, especially when they are treated as children, as in our society, is an important but moot question. While no systematic study has been made of the long-range consequences of the rate of maturation, it can be assumed that the fact of being early or late in development has some effect on self-confidence and on the ease of finding one's place in the social order of things. In this regard, it is paradoxical that our society increasingly tends to deny young men and women the chance of engaging in responsible activities, although their physiological and sexual development is accelerated.

The extensive studies in animals and man of the effects of nutrition on physical and mental development have made clear that many traits that used to be regarded as racially determined are in reality the consequences of economic factors and ways of life. The traditional small stature and thinness of the Japanese, as we have seen, gave way within one generation to a much larger body size among children brought up under conditions similar to those that prevail in the Western world. Similarly, the indolence and lethargy which used to be regarded as inherent in the racial make-up of Central American people appears in reality to be the result of nutritional deficiencies and of infections contracted during early life. This does not mean that the differences between population groups are all due to environmental factors. Genetic constitution certainly accounts in large part for the striking contrast between the very short size of pygmies in the Congo and the tallness of Dinka people in the Sudan. The genetic endowment may also account for some of the differences in behavior and mental attributes among various races. But granted the importance of genetic constraints, conditions of early life are certainly responsible for many differences that have long been assumed to be racial in origin.

Nutritional deficiencies and infectious disease are the

two groups of environmental factors that have been most extensively studied with regard to their effect on the physical and mental development of children and adults, but many other conditions of early life can also have lasting effects.

Chickens have been conditioned while still in the egg to certain visual and auditory stimuli; after hatching, their response to these stimuli is different from that of unconditioned animals.

More directly relevant to human life perhaps is the observation that if rats and mice are transferred to foster parents immediately after birth, their behavior as adults is profoundly and lastingly affected. Mice reared with rats by foster rat mothers even tend to prefer the society of rats to that of mice.

Students of animal behavior have described many other situations in which the normal ways of a species have been drastically changed by abnormal early experiences. For example, cats have been made to cohabit and enter a benign relationship with rats and mice by arranging that they be raised from very early life in constant association with these rodents. In one particular experiment, 18 cats were reared with rats in the same cages; when they reached adult age none ever attacked their cage mates and only 3 out of 18 ever killed a strange rat. Monkeys raised in unusual environments, some of which were extremely abnormal, such as complete isolation, later in life selected by preference the company of animals that had experienced the same type of environmental stress.[29]

Isolation can be as traumatic an experience for young animals as it is for human infants. This has been repeatedly demonstrated in dogs, but the most dramatic and detailed studies of the effects of early isolation on behavior have been carried out with primates deprived during early life of contact with their mothers or other young primates. At the Primate Center in Madison, Wisconsin, Professor H. F. Harlow and his colleagues demonstrated that rhesus monkeys removed from their mothers immediately after birth and raised under a variety of isolated conditions exhibited throughout their lives abnormalities in perception, learning ability, and sexual behavior. Even more spectacular and lasting disturbances in behavior were brought about when the monkeys were isolated from other infant monkeys during the first six months of their lives; as adults they were incapable of engaging in normal social play or sexual relations with other adult monkeys. In Professor Harlow's words, "They exhibit abnormalities of behavior rarely seen in animals born in the wild. They sit in their cages and stare fixedly into space, circle their cages in a repetitively stereotyped manner, and clasp their heads in their hands or arms and rock for long periods of time. . . . The animal may chew and tear at its body until it bleeds."[30] The first six months of the primate's life were critical in Professor Harlow's experiments. This period corresponds approximately to the first few years of a child's life. There is in fact very strong evidence that the extent and type of contact among children 3 to 10 years of age can greatly influence their subsequent emotional life.[31]

The harm done to children by isolation and other forms of social and emotional deprivation is now well recognized. Much less is known, however, of the effects of crowding and excessive social stimulation. Many experiments with various animal species have revealed that crowding commonly results in disturbances of endocrine function and of behavior. But the precise effects differ profoundly depending upon the conditions under which crowding is achieved. If too many animals are brought together in the same area after they have reached adulthood, they exhibit extremely aggressive behavior and a large percentage of them die. However, animals which are born within a given enclosure and allowed to multiply in it can live in very high densities of population without displaying destructive aggressiveness, because they achieve a social organization that minimizes violent conflict. As the population pressure increases, however, more and more animals exhibit abnormal behavior of various kinds. These deviants are not sick organically, but they act as if they were unaware of the presence of their cage mates. Their behavior is asocial rather than antisocial.[32]

It is true that men are not rats. But the most unpleasant thing about rats in crowded conditions is that they behave so much like many people in crowded human communities. Man has developed a variety of social mechanisms that enable him to live in densely populated areas. Holland, for example, is one of the most densely populated countries in the world, yet its people have excellent physical health and a low crime rate. In other communities, however, crowding may lead to types of asocial behavior that recall the social unawareness observed in overcrowded rodent populations.

The humanness of man is the product of a socialization process that begins very early in life; some of the most important aspects of personality develop even before the school years. The newborn human today is no different from what he would have been if he had been born thousands of years ago in a Stone Age culture. But he immediately confronts his twentieth-century mother, or nurse, and is subjected by her to a process of acculturation that inhibits some of his potentialities and lets others become fully expressed. The infant is thus shaped according to our socially accepted norms and adapts more or less successfully to our world. Some of the most "human" traits disappear in populations that are extremely crowded, probably because all human beings, and especially children, must make their contacts with other human beings under proper conditions in order to develop the potentialities best suited to the kind of life we desire. Man needs the socializing effect of a normal human group in order to become and remain normally human.

History shows that sudden increases in population density can be as harmful to man as they are to animals. The biological and social disturbances created by the intense crowding in tenements and factories during the nineteenth-century Industrial Revolution were probably most severe in groups that had immigrated from rural areas and were therefore not adapted to urban life. These disturbances are now becoming milder, even though the world is more and more urbanized and industrialized. Constant and intimate contact with hordes of human beings has come to constitute the normal way of life, and men have adjusted to it because a crowded environment is now commonly part of their early experience.

The most important effects of the early environment may be the ones that convert the child's inherited potentialities into the traits that constitute his personality. In this regard, it must be emphasized that mere exposure to a stimulus is not sufficient to affect physical and mental development. The forces of the environment act as formative influences only when they evoke creative responses from the organism.

In man as in animals, the physical and mental structure can be deeply affected only while the processes of anatomical and physiological organization are actively going on; the biological system becomes increasingly resistant to change after it has completed its organization. These statements are valid not only for anatomical and physiological differentiation, but also for the emergence of tastes, social attitudes, and even the perception of space in interpersonal encounters. Suffice it here to quote from an unpublished lecture by E. T. Hall, an American social anthropologist who has emphasized in several books[33] that people brought up in different cultures live in different perceptual worlds:

"Consider for a moment the difference between a Greek who garners information from the way people use their eyes and look at him, and the Navajo Indian whose eyes must never meet those of another person. Or consider the disparity between a German who must screen both sight and sound in order to have privacy, and the Italian who is involved with people visually or auditorily almost twenty-four hours a day. Compare the sensory world of the New England American, who must stay out of other people's olfactory range and who avoids breathing on anyone, and the Arab who has great difficulty interacting with others in any situation where he is not warmly wrapped in the olfactory cloud of his companion. All the senses are involved in the perception of space; there is auditory, tactile, kinesthetic, and even thermal space. . . .

"The kind of private and public spaces that should be created for people in towns and cities depends upon their position on the involvement scale."

The national differences in perception of space during interpersonal encounters are not racially determined; they are expressions of social influences rooted in history and experienced during early life. These influences affect also the perception of other aspects of the environment. Murky skies, ill-smelling air, noisy streets, vulgar design, uncouth behavior are accepted without protest and indeed remain unnoticed by those who have experienced these conditions from early life on. Paradoxically, the most frightening aspect of human life is that man can become adapted to almost anything, even to conditions that will inevitably destroy the very values that have given mankind its uniqueness.

Experiments with animals, and a few observations on man, indicate that there is a basis of truth in the old wives' tales concerning the effects of the pregnant woman's emotional experiences on some of the characteristics of her child.[34] Many types of stress occurring during pregnancy leave their mark on the unborn child by stimulating the secretion of hormones that migrate across the placental barrier. It has been shown in rats that hormones of the sexual, thyroid, and adrenal glands of the mother have a direct action on the central nervous system of the young and, if they act at a critical time, produce permanent effects on psychophysiological processes. When the male hormone testosterone is injected into pregnant monkeys, the behavior of their female offspring is profoundly altered; although anatomically female, these young animals display activities similar to those of the male offspring. Like the latter they engage in rough and tumble play, and they show little tendency to withdraw from the threats and approaches of others.[35]

It would be surprising if hormonal influences originating from the human mother did not have similar effects on brain organization in the child. It may be literally true that, in the words of an ancient Chinese philosopher, "The most significant period of an individual's life is spent in his mother's womb."[36]

Prenatal, neonatal, and other early influences thus constitute a continuous spectrum through which the environment conditions the whole future of the developing organism. The rates of physical or sexual maturation and the final adult size are not the only, or the most important, effects of these early influences. Physiological characteristics, tastes, interests, and social attitudes are also shaped early in life by environmental facts. Physically and mentally, individually and socially, the responses of human beings to the conditions of the present are always conditioned by the biological remembrance of things past. William Wordsworth's statement in "The Rainbow" that "the Child is Father of the Man" is a poetical expression of a broadly conceived biological Freudianism.

Of Human Nature

In common usage, the phrase "human nature" refers chiefly if not exclusively to the psychological and moral attributes of man. When used by biological scientists, the phrase denotes, in addition, the anatomical structures and physiological attributes of the human body, both the inherited ones and those that are acquired or modified by experience. Whether used in its limited or generalized sense, the phrase human nature has long been the subject of philosophical and scientific arguments identified with the nature versus nurture controversy.

The view that man is the product of his environment, so forcefully stated by Hippocrates in *Airs, Waters, and Places*, has long remained influential not only among physicians but even more among philosophers. John Locke (1632–

1704), Jean Jacques Rousseau (1712–1778), and other partisans of the "nurture" theory of human development believed that the newborn child is like a blank page on which everything is consecutively written in the course of life by experience and learning. In the spirit of this general theory, Rousseau's contemporary Claude Helvetius asserted that, intellectually, man is but a product of his education; Charles Fourier (1772–1837) went so far as to state that universities could at will produce nations of Shakespeares and Newtons! A century ago, Thomas Huxley (1825–1895) asserted with his usual picturesque vigor that the newborn infant does not come into the world labeled scavenger or shopkeeper or bishop or duke; he is born as a mass of rather undifferentiated red pulp and it is only by educating him that we can discover his capabilities.

In contrast to the partisans of "nurture," Thomas Hobbes (1588–1679), Herbert Spencer (1820–1903) and the social Darwinists upheld the view that nature (heredity) determines to a very large extent the characteristics of the person, young or adult. On the basis of very inadequate statistical evidence, Francis Galton (1822–1911) concluded that this genetic view accounted satisfactorily for the stratification of English society. As he saw it, judges begot judges, whereas workmen, artisans, and even businessmen were not likely to be born with the innate mental ability required for a successful performance in the intellectual world. From Joseph-Arthur Gobineau (1816–1882) to Adolf Hitler (1889–1945) and into the present, a narrow interpretation of genetic determinism has given rise to many foolish and criminal attitudes concerning the existence of inferior and master races.

The conflict between genetic and environmental philosophies in the analysis of human attributes has continued into the twentieth century. Sigmund Freud (1856–1939) believed that the peculiarities of each person's mind can be accounted for by the influences that have impinged on his development, especially those around the time of birth. According to Freud and his followers, most of the complexes that plague man's existence are determined by the early environment. In contrast, Carl Jung (1875–1961) claimed that man can be understood only by exploring the many factors which played a part in the genesis of the collective human mind during the remote past. He related behavior to the operation of archetypes as old as the human race itself.

Modern discussions concerning the development of languages also center on the genetic-environmental theme. At Massachusetts Institute of Technology, Professor Noam Chomsky teaches that there is an intuitive semantics common to the human species and underlying all spoken languages.[37] In contrast, the Swiss child psychologist Jean Piaget doubts that this universal grammar is really innate; he points out that speech ability is not present at birth and that speech does not become possible until the major sensory-motor functions have become organized to the point where they are capable of generalization. In Piaget's view, man's universal semantic ability might depend upon the fact that all human beings have similar experiences in early life, leading to the organization and interrelation of the sensory-motor systems.[38]

The nature versus nurture controversy constitutes only a pseudo problem, because, as stated earlier, genes do not determine the characteristics by which we know a person; they merely govern the responses to experiences from which the personality is built. Recent discoveries are beginning to throw light on the mechanisms through which environmental stimuli determine which parts of the genetic endowment are repressed and which parts are activated.

Microscopic and chemical studies have revealed the remarkable fact that at any given time in any specialized cell only a limited number of genes are active—10 to 15 percent of the total gene areas is probably a reasonable figure. This is true of nerve cells as well as of any other type of differentiated cells. Furthermore, genes can be activated or repressed by certain kinds of substances, hormones in particular.[39] It can be assumed that gene activity is profoundly influenced by the composition of the cellular fluids and that various substances differ qualitatively and quantitatively in their activating or repressing effects.

A general hypothesis can now be formulated to account for the well-established fact that the external environment conditions the manner in which the genetic endowment of each person becomes converted into his individual reality. This hypothesis states that the external environment constantly affects the composition of the body fluids, in part by introducing certain substances directly into the system, in part by affecting hormone secretion and other metabolic activities. Such changes in the body fluids alter the intracellular medium which in turn affects the activity of the genetic apparatus. In this manner, the individual's experiences determine the extent to which the genetic endowment is converted into the functional attributes that make the person become what he is and behave as he does.

The biological and psychological uniqueness of every human being generates many conceptual and practical difficulties for the study of behavior and the practice of medicine.

Theoretical scientists can make generalizations about man's biological nature, but psychologists and physicians must deal with individual persons. Each person represents a constellation of characteristics and has problems that differ from those found in any other person.

Until a few decades ago, physicians commonly referred to a patient's "constitution" when discussing problems of diagnosis and prognosis. They used the word constitution to denote the person's physical and mental characteristics relevant to his state of health. The patient's constitution was assumed to determine his susceptibility and resistance to stresses and to trauma, as well as his ability to overcome the effects of disease. The term is now rarely if ever used in scientific medicine, because its meaning seems extremely vague. This is regrettable, because much progress has been made toward understanding the factors that determine how a given person will respond to a biological or psychological threat. One definition of the word constitution in Webster's Third International Dictionary is "the whole physical makeup of the individual comprising inherited qualities as modified by the environment."

Early influences certainly play the most important role in converting the genetic potentialities into physical and mental attributes, but it is obvious that these attributes change continuously throughout life. The changes occur as a result of the aging process, and also because physical and mental attributes are constantly being acted on, and thereby altered, by environmental stimuli. To live is to function and to respond. Almost every response of the organism to any stimulus results in the acquisition of memories that alter its subsequent response to the same stimulus. Two organs of memory are now recognized—the brain and the so-called reticulo-endothelial system, which provides the mechanism for a sort of biological memory.

The brain is able to register and store experiences until the time of death. Whether the memory is conscious or unconscious is immaterial for the present discussion. The point of importance is that even subconscious memories can be activated, either by occurrences with which the past events were associated, or by artificially stimulating the proper area of the brain.

The reticulo-endothelial system is a complex of cells widely distributed throughout the body which can bring about tissue changes resulting in the various forms of immunity and allergy. For example, human beings are never spontaneously sensitive to poison ivy; they become allergic to it only after having been exposed. When allergic sensitization has occurred the sensitized person retains the allergy long after the sensitizing event. Human beings susceptible to tetanus toxin can acquire antitoxic immunity by the proper technique of vaccination with this toxin or with a detoxified derivative of it; immunity may wane with time, but some evidence of it persists for many years and probably for the whole life span. Allergy to poison ivy or to any other substance, and immunity to tetanus toxin or to any other poison or microbe, can be regarded as manifestations of biological memory.

Each person's constitution is therefore made up of the evolutionary past embodied in the genetic apparatus and of the experiential past incorporated in the various forms of mental and biological memory. Throughout life, the constitution becomes modified and enriched by the responses that the body and the mind make to environmental stimuli and that become incorporated in the physical and mental being of the person—incarnated in his being, so to speak. At any given time, the constitution of a particular person includes the potentialities that his experiences have made functional; its limits are determined by his genetic endowment. Since the constitution changes continuously with time, it can be defined in scientific jargon as the continuously evolving phenotype of each particular person.[40]

With the childlike confidence in the power of the intellect characteristic of his time, Francis Bacon (1561–1626) wrote: "A man is but what he knoweth." Even if the meaning of this assertion were enlarged to include all aspects of physical and mental awareness, it would still fall short of describing the human condition, because many attitudes and responses have determinants of which neither the mind nor the body is aware. Blaise Pascal (1623–1662) sounded a note much more congenial to the modern mood when he wrote in his *Pensées*: "*Le coeur a ses raisons que la raison ne connait pas.*" Despite its obscurity, this phrase conveys a universally valid experience: the heart does indeed have its own reasons which reason does not know.

The inner life of modern man is still influenced by very ancient biological processes that have not changed since Paleolithic times and of which he is often completely unaware. The heart to which Pascal referred includes all the determinants of behavior that do not originate from conscious reason and that often escape its control. In addition to the determinants that survive from man's evolutionary past and are common to all mankind, there are those acquired by each person in the course of his own individual life. Such acquired determinants of physiological responses and mental attitudes naturally differ from culture to culture and from person to person.

In most situations, human behavior includes spontaneous nonrational processes originating from a part of the brain (the diencephalon or thalamoencephalon) that developed during evolution long before the more superficial cortex. Terror, rage, the various forms of animal desire—indeed all the so-called lower impulses—involve instinctive processes that have intense physiological manifestations in the organism even if they are not expressed in overt behavior. Social life in the most primitive as well as the most highly evolved societies requires that the spontaneous impulses originating from the diencephalon be controlled by the higher so-called rational processes, monitored through the cerebral cortex, and related to the culture of the particular society. The expression "cortical conceit" has been coined to denote the belief that man often ignores the ancient evolutionary components of his nature and lets his behavior be completely ruled by directives that are culture-inspired and originate from the cerebral cortex. Many of the biological difficulties peculiar to mankind have their origin in this so-called cortical conceit.[41]

Philosophers, writers, and artists have always acknowledged explicitly or tacitly the immense role played by these ancient and unconscious biological processes. Plato (427–347 B.C.), in the dialogue *Phaedrus,* has Socrates speak with passion of the creative forces released in man by "mania" or the "divine madness."[42] The text of the dialogue makes it clear that the word "madness" as used by Plato refers not to a diseased mental state, but rather to the deep biological attributes of man's nature which are almost beyond the control of reason and often transcend its reach. In the usual circumstances of ordinary life these attributes remain concealed, but under certain situations they constitute inescapable imperatives; more interestingly, perhaps, they can become powerful sources of inspiration for the creative individual. Ethical attitudes and intellectual creativity depend in part on the ability to hear "the voice of the deep" and to tap resources from regions of man's nature which have not yet been explored.

Remarkably, Descartes, the innovator and most ardent advocate of pure reason, has recounted that he discovered his famous "method" when very young during an ecstatic vision that he always regarded as the culminating moment of his life, something in which he barely had a role, a divine gift, a transcendental revelation. He inscribed in his personal notes: "X *novembrix 1619, cum plenus forem. Enthousiasmo, et mirabilis scientiae fundamenta reperirem*" ["10 November 1619, when I was full of enthusiasm, and I discovered the fundamental principles of a wonderful knowledge."][43]

Friedrich Nietzsche (1844–1900) was referring to innate forces analogous to Socrates' divine madness when he wrote in *The Birth of Tragedy* that the Dionysian riotous spontaneity is a necessary complement of the Apollonian rational order. In fact, as shown by the English scholar E. R. Dodds, ancient civilizations were aware of powerful biological needs of man's nature which are not clearly perceived and thus appear irrational.[44] They symbolized such occult passions—the divine madness—by a ferocious bull struggling against reason. Since these forces operate independently of reason rather than against it, they should be called nonrational rather than irrational.

Empirically, all over the world, social practices have been developed to let nonrational forces manifest themselves under somewhat controlled conditions. Among the Greeks, the Dionysian celebrations, the Eleusinian mysteries, and many other rituals served as release mechanisms for biological urges which could not find an otherwise acceptable expression in the rational aspects of Greek life; even Socrates participated in the Corybantic rites. Such ancient traditions still persist even in the most advanced countries of the Western world, though often in a distorted form; they extend all the way from New Year celebrations to Maypole dances, from the orgies of Mardi Gras to the rhythmic excitement of rock and roll. Even in the urbane city dweller, the Paleolithic bull survives and paws the earth whenever a threatening gesture is made on the social scene. The passions depicted by classical tragedies have biological roots deep in the Paleolithic past.

In animals, including the higher primates, most behavior is instinctive and intellectually neutral. It is rarely if ever oriented toward a distant future that the animal tries to predict and to bring about willfully. In contrast, man's responses to most environmental stimuli are profoundly affected by anticipations of the future, whether these anticipations are based on fear, factual knowledge, desire for achievement, or merely wishful thinking. Indeed, man's propensity to imagine that which does not yet exist, or would never come to pass without his willful and deliberate action, is the aspect of his nature that differentiates him most clearly from animals. It also contributes greatly to the complexities of his "constitution" that so baffle physicians.

One of the most distinctive aspects of human life is the tendency to transcend simple biological urges; man is prone to convert ordinary processes of existence into actions, representations, and aspirations that have no biological necessity and may even be inimical to life. Furthermore, he tends to symbolize everything that happens to him and then to react to the symbols as if they were real external stimuli.[45] The response of any given person to an environmental factor is conditioned both physiologically and psychologically by his own past experiences; it is therefore highly personal. The power of the personal past is so great that it can distort the meaning of any event and magnify trivial happenings into momentous experiences. Human reactions are so profoundly influenced by the individual past that they are usually unpredictable and therefore appear completely irrational.

Through complex mechanisms that are only beginning to be understood, all the perceptions and apprehensions of the mind become translated into organic processes. The body responds not only to the stimulus itself, but also to all the symbols associated with the memories of the past, the experiences of the present, and the anticipations of the future. Anything that impinges on man thus affects both his mind and his body and causes them to interact—an inescapable consequence of the evolutionary and experiential past.

In the course of human evolution, the brain, the body, and culture developed simultaneously under one another's influence, through the operation of complex feedback processes.[46] Integrated interrelationships of biological constitution and of function necessarily resulted from this evolutionary interdependence of body, brain, and culture. The activities of the human brain imply certain characteristics of the body and certain patterns of culture.

Like the evolutionary development of mankind, the experiential development of each individual person consists in an integrated series of responses to environmental stimuli. In most situations, the effects of the total environment on the body and the mind must therefore be interrelated, because exposure to almost any kind of stimulus must evoke into activity patterns of physical and mental responses that are associated because they were simultaneously established by past experiences. The view that the bodily and mental constitution consists in the biological memory—genetic and experiential—of interrelated responses made in the past provides a theoretical basis for psychosomatic medicine. More generally, it accounts for the fact that human nature, in health and in disease, is the historical expression of the adaptive responses made by man during his evolutionary past and his individual life.

Genetic and experiential factors interplay to shape the biological and behavioral manifestations of human life, but they do not suffice to account for the totality of human nature. Man also enjoys a great degree of freedom in making decisions; he is *par excellence* the creature that can choose, eliminate, organize, and thereby create. Human nature will not be fully understood until it becomes possible to relate its two complementary aspects—determinism and free will. All social practices and ethical attitudes are based on physiological needs, urges, and limitations woven in the human fabric during the evolutionary and experiential past. But within the constraints imposed by the biological determinants of his nature, man can make responsible choices. He has the privilege and the responsibility of shaping his self and his future.

4.

THE LIVING EXPERIENCE

Man Umbilical to Earth

For a particular organism the environment that is meaningful is the world of colors, tastes, and sounds that it perceives, the irritating, stimulating, soothing, and repressing influences that affect its physiological and mental being. Animals, and especially men, do not merely react as passive objects to the environment; they shut out certain aspects of it and select others to which they respond in a personal and often creative manner. This highly personal interplay between a particular organism and its environment constitutes what I shall call the living experience.

The Danish physicist Niels Bohr (1885–1962) did not believe that science could achieve an objective description or explanation of reality. What scientists really try to do, Bohr asserted, is to develop ways of stating without ambiguity the *experience* they gain of the world, either directly by observation or indirectly by instrumentation and computation.[1] The word "reality" thus has large subjective components because it involves the nature of the personal experience.

The language of physicists is highly metaphorical in its use of such phrases as "elementary particles" or "electron orbits"; physicists have never seen either the particles or the orbits. When they describe the hydrogen atom, they are not referring to an object which has reality for non-physicists. Mathematical concepts such as wave functions cannot have any relevance to the real world of the man in the street; they correspond to another kind of reality experienced indirectly through suitable measurements meaningful only to a small number of specialists.

Biologists, similarly, give their own specialized meaning to the word reality when they deal with structures or functions of living organisms which cannot be perceived directly by the senses. When they first used the expression "gene" a few decades ago, they had in mind a picture far different from that generally conveyed by this word today; moreover, the picture will certainly continue to change with new methods of investigation. As to the mechanisms postulated to account for vision or learning, they reflect at each period some contemporary aspects of physicochemical theory but have little bearing on the experiences of the viewer or the learner.

If the word "real" can be used at all, according to the poet W. H. Auden, it must be with reference to the only world which is real for us. "The world in which all of us, including scientists, are born, work, love, hate and die, is the primary phenomenal world as it is and always has been presented to us through our senses, a world in which the sun moves across the sky from east to west, the stars are hung in the vault of heaven, the measure of magnitude is the human body, and objects are either in motion or at rest."[2]

Since the real world of which Auden speaks is perceived naturally through the senses, man's awareness of it can be studied scientifically by analyzing the mechanisms through which the body registers environmental stimuli. As far as the perceiving organism is concerned, however, what really matters is its experience of the total environment, rather than the processes through which it apprehends reality.

The past, as we have seen, plays a large role in determining the manner in which man experiences the environment. Heredity and memory are two different mechanisms through which the stimuli that impinge on the body and mind leave on them permanent traces or imprints known as engrams. Since all the subsequent reactions and responses of the organism are conditioned by such engrams, these probably act as directive agencies for further development.

The act of bringing memory traces to the surface is just as important an activity as the storage of information. The past affects the fate of the organism not because it has been stored, but because many stimuli bring about its retrieval and thus condition all physiological and behavioral responses. For example, many instinctive actions which are not prompted by immediate needs have significance for the subsequent development of the organism or for the survival of the species. The manner in which we experience the present is still further complicated by our hopes and conscious anticipations of the future. In fact, conscious concern for the future is one of the most important attributes distinguishing man from animals. But this does not mean that explicit and logical thought is the most influential determinant of human behavior.

Logical thinking has now been incorporated into so-called thinking machines which are highly effective in carrying out operations once assumed to be the prerogative of the human brain. Herbert A. Simon, professor of computer science and psychology at the Carnegie Institute of Technology, recently pointed out that "the capacities we are having real trouble getting machines to simulate are not the higher order of faculties confined to man, but rather the ones man shares with the lower animals."[3] These primitive attributes are of enormous importance in human life.

Ever since the late nineteenth century, the keynote of most biological and social thinking has been the evolutionary process: biological evolution occurring through the selection of mutants by environmental factors, and psychosocial evolution occurring through cultural agencies. Equally essential for the understanding of human life, however, is the recognition that certain aspects of man's nature are almost unchangeable in a timeless now—the eternal present.[4]

Just as the dog and the cat, though they have been domesticated and pampered for thousands of years, still retain the fundamental characteristics of their wild ancestors, so does modern man exhibit many traits that have survived from his distant past. Many of his present activities are derived from very ancient ways of life culturally transmitted from one generation to the next, even though often in a highly distorted form. For example, the celebration of Carnival can be traced to fundamental impulses of man's nature that are incompatible with life in complex societies yet demand to be admitted now and then. For a few days during Carnival, the forces of the irrational are allowed to take control over law and order, symbolically at least. In Catholic countries the culminating day of this strange interlude in men's lives is Mardi Gras—Fat Tuesday—a day of symbolic license permitted the faithful on the eve of Lenten austerities.[5] Similarly, as we have seen, many other aspects of modern life are still influenced by attitudes and practices the origin of which can be traced as far back as the beginnings of the Sumerian civilization many thousand years ago.[6]

Certain gestures and sounds also retain a deep hold on man. Whether the power of their influence is of genetic or cultural origin is irrelevant here. The point of importance is that modern man is just as susceptible to them as were his distant ancestors—a fact that political and military leaders well know and constantly utilize. The comments of Han Suyin, a Chinese woman writer who in her book *A Many-Splendored Thing* described the atmosphere in China at the end of the Civil War, are applicable to people of other countries and other races, irrespective of social institutions: "I wonder whether . . . our ancestors held their Spring Festival and their Fertility Rites to this dancing and this beat? It is from deep within our people, this bewitchment of drum and body. I feel it surge up from my belly, where all true feeling lies; strong and compelling as love, as if the marrow of my bones had heard it millions of days before this day."[7]

The drives to explore the environment, to delimit a territory, and to become familiar with a home range are among the most fundamental aspects of animal behavior.[8]

Such exploratory activities have much to do with what is generally called play, but they constitute in reality an effective manner of establishing through experience a close relationship with the outer world. People in primitive tribes also explore their environment and thus acquire a deep knowledge of its resources, and its dangers. Even in the most civilized and technicized societies, play remains essential for the acquisition of knowledge, especially for self-discovery by the child and the adolescent. The drive to explore and to play probably contributes also to the continued growth of the adult. It may well be true that, as the sculptor Constantin Brancusi (1876–1957) is reported to have said, when we are no longer young we are already dead.

All students of primitive life have noted that the senses of human beings who live close to nature are much keener than those of civilized man. As is well known, the Paleolithic painters performed feats of animal representation surpassing the achievements of even the most accomplished animal draftsmen of our times. The English art critic Roger Fry (1866–1934) pointed out that, in Paleolithic representation of trotting animals, "the gesture is seen by us to be true only because our slow and imperfect vision has been helped out by instantaneous photography. Fifty years ago we should have rejected such a rendering as absurd."[9]

Modern man retains the same potentialities for keenness of perception that his distant ancestors had, as demonstrated by the fact that persons who have removed themselves from technicized environments commonly display increased ability to perceive colors, sounds, and odors. In his book *A Hind in Richmond Park*, the English naturalist W. H. Hudson (1841–1922) reported spectacular examples of acuteness in the perceptions of smells by South American Indians, and by Europeans returning from a prolonged sea voyage or sojourn in the mountains.[10] Similarly, a marked increase in sharpness of vision was experienced recently by a group of young Frenchmen when they returned to the earth's surface after spending several months in a deep, dimly lighted cave.[11]

In the ordinary practice of civilized life, modern man feels compelled to confine himself largely to the stimuli that are germane to his purpose of controlling the environment. By using his consciousness to avoid the impact of many stimuli that stream out of total reality he certainly simplifies his life and increases his efficiency, but at the cost of much impoverishment. The success on American college campuses of Norman Brown's *Life Against Death*[12] probably indicates that many young people are aware of the impoverishment of physical and emotional life that results from the atrophy of sense perceptions brought about by present-day existence. They are probably right in believing with Brown that "the resurrection of the body" is an essential condition of mental sanity.

The widespread acceptance of the nonreligious attitude in modern Western societies has placed agnostic man in a difficult situation. Although he has carried to the extreme the desacralization process he cannot free himself entirely from the past. His ancient religious nature always persists in his deepest being, ready to be reactivated, because he is haunted by the very realities he tries to deny.[13]

New Year celebrations, housewarming ceremonies, marriage rites, or parties at the birth of a child commonly take the form of camouflaged and degenerated myths. The myths and rituals of modern man constantly come to light in the plays he patronizes, the magazines and books he reads, the dreamlands depicted on moving-picture and television screens. The fights between heroes and monsters, the paradisiacal landscapes and descents into hell are themes of modern storytelling just as popular as they were in the ancient or medieval worlds. Reading books and watching spectacles fulfill a mythological function, enabling man to escape from time as he did in the past through his ancient rituals.[4,13]

Simple words such as air, water, soil, and fire evoke in most human beings deep emotions that recall the past; just as they did for primitive people, these words stand for the very essence of the material creation. Despite much knowledge of the physical and chemical properties they represent, they still convey to most people a sense of eternal and essentially irreducible value. Fire, for example, remains a great reality with mystic undertones, probably because human life has been organized around it for ages.[14] Fire as a concept has progressively disappeared from science during the past few decades; the chapters devoted to it in textbooks of physics and chemistry are becoming shorter and shorter, when they exist at all. But the words flame and fire remain just as deeply meaningful for real human life—including the life of physicists and chemists.

Whatever science may have to say about the fundamental processes and constituents of the natural world, we regard Nature holistically and respond to it with our whole physical and emotional being. Deep in our hearts we still personalize natural forces and for this reason experience guilt at their desecration. The manifestations of Nature are identified with unchangeable needs of human life and are charged with primeval emotions because man is still of the earth earthy.

In an attempt to be objective, scientists reduce the *I* to an abstract knowing subject and the *It* to the passive and abstract object of thought. According to the Jewish theologian Martin Buber (1878–1965), this impersonal I-It attitude does not permit dealing scientifically with the wholeness of man. Objective science is limited to the study of selected aspects of man's nature, considered as ordinary parts of the natural world. Studying the species *Homo sapiens* or individual men in comparison with animal species or with other individual men can only reveal similarities and differences, not man as a whole. It categorizes men and animals as differing objects but does not recognize the uniqueness of man which resides in the I-Thou relation.

Buber's I-Thou philosophy asserts that the essence of man is the experience of his relation to other human beings, and to the cosmos. Because of the subtlety—or obscurity—of Buber's views, it seems best to use his own words. He defines man as "the creature capable of entering into living relation with the world and things, with men both as individuals and as the many, and with the mystery of being which is dimly apparent through all this but infinitely transcends it." The uniqueness of man is to be found not in the individual, or in the collective, but in the meeting of "I" and "Thou." "The fundamental fact of human existence is neither the individual as such nor the aggregate as such. Each, considered by itself, is a mighty abstraction. The individual is a fact of existence in so far as he steps into a living relation with other individuals. The aggregate is a fact of existence in so far as it is built up of living units of relation. . . . That essence of man which is special to him can be directly known only in a living relation."[15]

The real I-Thou encounter as defined by Buber is a

rare experience, but other types of human relationships even though less profound are nevertheless desirable.

In *The Secular City*, the theologian Harvey Cox asks: "Besides I-It relationships, in which the other person is reduced to the status of an object, and in addition to the profound, personally formative I-Thou encounter, why could we not evolve a theology of the I-You relationship? . . . It would include all those public relationships we so enjoy in the city but which we do not allow to develop into private ones."[16] Jane Jacobs had the same thought in mind when she wrote: "Cities are full of people with whom, from your viewpoint, or mine, or any other individual's, a certain degree of contact is useful or enjoyable; but you do not want them in your hair. And they do not want you in theirs either."[17]

For the laboratory scientist, life consists in the set of integrated processes that keep organisms functioning and that enable them to reproduce themselves and to evolve. But human beings experience life much more richly by relating their whole self to other human beings and to the cosmic order. One need not understand all the philosophical complexities of the I-Thou relationship to realize that man is organically bound to social structures. One need not be a Christian, or even have any religious belief, to experience elation at hearing the bells of a cathedral on Easter morning. Bells resounding in the skies symbolize man's integration with the cosmos. Their sounds, endlessly modified as they spread in all directions, reach into the far and beyond, where, despite space explorations, all is mystery. Even the most skeptical and sophisticated modern man recaptures once more the experience of being part of the total cosmic order as he follows the voices of the bells dispersing themselves into the sky. Experiencing the Easter bells makes man aware that his life demands obedience to universal laws.

One of the most ancient and continuing preoccupations of mankind has been the patient and subtle effort to understand man's relation to other men and to the rest of the world. There have been as many answers to the riddle as there are religions and philosophies.

At one extreme is the solution proposed in India by the sages and saints who formulated the Upanishads 2,500 years ago. They taught that a simple and essential reality underlies the superficial multiplicity of things and events; in their view, the essence of the self is not the body or the mind, or the individual ego, but the formless and silent being within each person—the Atman, identical with the impersonal Soul of the World. According to the Upanishads, the intellect is inadequate for reaching the ultimate significance of life. "As flowing rivers disappear into the sea, losing their name and form, thus a wise man, freed from name and form, goes to the divine person who is beyond all."[18]

The wistful theosophical concept that individual life is a delusion has dominated Hindu thought since the time of Buddha and has also enlisted many followers in Europe and America. But despite its advocacy by Arthur Schopenhauer (1788–1860) and Ralph Waldo Emerson, it has never proved congenial to Western religions and institutions, which are in contrast permeated with the cult of individualism. The most dominant characteristic of Western culture has been the search not for the essential unity that underlies existential multiplicity, but rather for ways to convert the universal characteristics of mankind into a great diversity of individual experience.

The Emergence of Individuality

There is a story about four young German artists who once set out to paint at the same time the same landscape in Tivoli near Rome. They were close friends, came from similar backgrounds, and had received the same training. They promised one another that they would copy the landscape as faithfully as possible so as to represent nature with accuracy and objectivity. Yet, as could have been expected, they produced four very different pictures. A painting, Jean-Baptiste Corot said, is nature seen through a temperament.

The adventure of the four German painters illustrates the fact that there is no such thing as objective vision and representation, because each person experiences the world and responds to it in his own particular way. This uniqueness of experience creates difficulties for the understanding of human behavior and accounts for many of the differences between the artistic and the scientific enterprise.

Artists and scientists deal with the same world, but they differ in their intellectual attitudes and in the techniques they use to recognize and describe objects, persons, and events. Artists focus their attention on private experiences, scientists on the generic aspects of nature. This difference in attitude is so fundamental that the aspects of the world with which science and art are respectively concerned have little in common. Even when they are looking at the same plant, animal, or person, artists and scientists become interested in entirely different manifestations of existence and think about different problems. The scientist wants to know the components and structures of which the living organism is made, the reactions which keep it alive, the effects that environmental forces exert on it. Since he regards knowledge bearing on the elemental structures, functions, and responses that are common to all forms of life as the most fundamental aspect of reality, he tends to minimize the differences between molecule, microbe, plant, animal, and man, and to select for investigation whatever organism or substance happens to be most suitable for the analytical problem he has in mind.

The artist, in contrast, is little if at all concerned with the elemental structures and mechanisms common to all living organisms. What he tries to perceive and to express is the living experience of individual creatures, in particular of individual men and women responding to the stimuli and challenges of the total environment in their unique way with all their complex attributes and aspirations.

The generic knowledge of *Homo sapiens* is not sufficient to account for the manner in which each individual person develops his own peculiarities, responds in his own way to environmental stimuli, and behaves as he does—in brief differs from all other human beings in his experience of the world. Yet the sense of discreteness and uniqueness is extremely pronounced in most persons, as evidenced by the fact that even the most reasonable man tends to be somewhat irritated when his name is misspelled or mispronounced. The failure of theoretical biology to deal meaningfully with the private experiences of each individual person accounts in large part for the widespread feeling that much of scientific knowledge has little relevance to the really important problems of human life.

One might assume that the attributes, needs, and aspirations which make up the whole man could be understood through a synthesis of humanistic and scientific knowledge. But this cannot be done, because the analytical aspects of human biology most extensively studied by scientists are different in kind from the phenomena of the living experience that artists try to express and humanists to comprehend. These two areas of knowledge do not form a coherent structure and indeed are almost unrelated.

Present scientific fashions notwithstanding, there is no valid reason for believing that the most important problems of life concern its analysis in terms of genes, their subunits, or the chemical reactions they control. Far more important, it might be argued, are the complex interrelationships between living things and their total environments. Civilizations are generated by such interrelationships. Life is so profoundly influenced by the evolutionary, experiential, and social past that even a highly sophisticated physicochemical approach leaves out of consideration most of its determinants and manifestations. "We murder to dissect," Wordsworth said in "The Tables Turned."

Since the living experience disappears when the organism is taken apart, many aspects of human life can be understood only by studying man's functioning with all its complexities and in the responses that he makes to significant stimuli. Such a study would require an organismic and ecologic attitude very different from the analytic one which now prevails in biology.

Admittedly, the humanness of man creates problems not definable in exact scientific terms. One may even question whether there does not exist an unbridgeable gap between objective knowledge of the generic aspects of mankind and the private experiences, to a large extent unsharable, of the individual person. Many human responses, fortunately, lend themselves to objective analysis.

The task is easiest for the responses which are directly elicited by environmental forces acting on the body machine. Physicochemical reactions and the reflex mechanisms of behavior can be studied by orthodox scientific methods, but such simple phenomena account for only a small proportion, and the least interesting, of the experiences that make up human life. In general, the stimuli that impinge on man set in motion a host of secondary processes which have complex indirect effects conditioned by past experience. Seeing a given object that brings to mind an article of food may stimulate appetite in one person and cause nausea in another; smelling an artificial perfume may evoke the heat of a summer day or the chill of a fall evening; hearing a faint but unexpected noise at night may cause the blood pressure either to rise or to fall.

Since the characteristics of physiological and mental responses are determined as much by the peculiarities of the person involved as by the nature of the stimulus, study of the living experience requires understanding of individuality.

The identity of each individual person is made up of characteristics that continuously undergo changes because the living expression of inherited potentialities is continuously being shaped by the conditions under which the person develops and functions. This cumulative quality of development is especially striking with regard to mental attributes. Children who acquire a rich stock of perceptual patterns and verbal labels early in life tend to have greater facility in building up the more complex patterns and labels required for conceptual thinking later in life.

The shaping of individuality by the environment was clearly formulated in the eighteenth century by the French philosopher Etienne Bonnet de Condillac (1715–1780). In his *Traité des Sensations,* Condillac created the fiction of a statue which was organized like a man but provided with a mind without ideas. By calling the various senses of his thoughtless statue-man into activity one after the other, he progressively endowed it with attention, memory, imagination, and reflection, showing that its needs, abilities, and ideas would be shaped by the environment in which it was placed.[19] One of Condillac's commentators, F. A. Lange, was even more explicit in suggesting that human beings acquire their characteristics through the impressions derived from their senses:

"Let us assume that in a faintly lighted subterranean chamber from which all sounds and sense impressions have been excluded, a newborn child is being scantily nourished by a naked and silent nurse, and that it is thus brought up until the age of twenty, thirty, or even forty, without any knowledge of the world or of human life. At this age, let him leave his solitude. And now let him be asked what thoughts he has had in his solitude, and how he has been nourished and brought up. He will make no answer; he will not even know that the sound addressed to him has any meaning. Where then is that immortal particle of divinity? Where is the soul that enters the body so learned and enlightened?"[20]

No one nowadays assumes that the soul enters the body to endow it with mental attributes, and it is universally accepted that the characteristics of each particular person have their basis in the genetic endowment acquired at conception. It is also recognized, on the other hand, that the human personality does not evolve merely from the unfolding of innate genetic traits according to a pre-established time sequence, but rather from the organization and differentiation of these potentialities by learning from experience. Sensations organize into patterns the activity which goes on spontaneously and continuously in nervous tissue.

The human baby does not at first distinguish between the subjective and objective world but learns to do so as his perceptions become organized.[21] His early life is largely devoted to the building up of an immense number of response patterns to stimuli of all sorts. The aspect of this pattern organization which is most typically human is that the vocalizings are shaped and integrated into the symbols used in speech. What we call the infant's personality is the outcome of an integration and compromise between his innate characteristics and the norms of social propriety and normality prevailing among the people who surround him.

Jean Piaget has attempted to dissociate the early adaptive processes of the child into two closely interrelated components—assimilation and accommodation. According to him, assimilation involves changes in the elements to which the child is trying to adapt—food or experience, for example; these elements can then be incorporated in the structure of the organism. Accommodation denotes modifications in the organism itself, in the digestive or mental systems, for example; these modifications enable the organism to adapt more successfully to the new situation.

Intellectual growth depends on active responses by the developing child. In order that information derived from the environment may become *formative* instead of being merely *informative*, the body and the mind must respond creatively to its impact. The structures elicited by such creative responses Piaget calls "schemas." The assimilation of new experiences increases the complexity of the child's schemas and this in turn enables him to achieve more complex accommodations. Moreover, still according to Piaget, the child's schemas do not necessarily remain unchanged if environmental stimulation is deficient; meanings are constantly being reorganized in his mind and linked with other meanings.[22]

The shaping of personality through responses to environmental stimuli has an anatomical basis in the brain's structure, because functional stimulation activates structural development in the nervous system. Anatomically, as well as intellectually, the brain develops with use and wastes away with disuse.[23]

During the evolutionary emergence of man, the enlargement of the neocortex was intimately related to the increasing complexity and changing character of social relationships. Human behavior has thus evolved from pre-existing patterns inscribed in the genetic code. To be functional, however, behavioral potentialities which are pre-established (nature) must be activated (nurture). When the human child reaches the age of 2½ to 4 months, for example, he can be caused to smile almost as readily by models of the human face as by a real face. Indeed a configuration made up of two eyes, a forehead, and some simple motion such as nodding or mouth movement, constitutes a sufficient stimulus for eliciting the smile response. Vision is the major mode of contact in the first half year of life, and smile the major source of rapport between parent and child. When we were babies the human face was the all-important source of welfare, and many actions were consequently related to it. No one can entirely escape this form of animism, because our brains have been shaped by this early experience.

The simple isolated response patterns of very early life become combined into larger units during late childhood and early adolescence; furthermore, development is profoundly influenced by social and historical forces. There was a time when children participated in practically all manifestations of adult life almost from infancy, but the modern family structure provides little chance for such solidarity and instead tends to break up the continuity of tradition.[24] This change may account for the fact that the rift between childhood and adulthood has continuously widened during the past two centuries, making approach to maturity increasingly difficult.

What we perceive and respond to constitutes the world we factually inhabit. A songbird lives very largely in a world of sound and vision, whereas a dog lives more in a world of smells. Man may differ from animals in being more independent of external impressions, but that this independence is not absolute is shown by the fact that his mental equilibrium and intellectual abilities rapidly deteriorate when his senses are kept inactive.[25]

The way the organism experiences the external world and responds to its stimuli must naturally be compatible with its own survival and the survival of the species to which it belongs. In fact, sensory receptors and their central connections in the nervous system possess structural and functional attributes that enable the organism to detect, and respond to, those aspects of the environment that are most essential for effective functioning.[26]

For example, vision receptors are tuned to wavelengths that are common in the environment in which the organism spends its life; the nervous retinal layers and the brain combine this information into functional structures. Similarly, the auditory receptors are tuned to frequencies that have adaptive significance for the species. The ability to sense sounds at 50,000 cycles per second contributes to the detection of small obstacles in the flight of a bat. Man, however, does not need to hear such sounds; his hearing mechanisms are adapted to vibrations of the order of a few thousand cycles.

This exquisite adjustment of sensory and brain mechanisms to crucial environmental features has its origin in the selective processes that shaped the species during the evolutionary past; for each individual organism it comes into being through the nervous connections laid down according to certain specific patterns under the influence of the stimuli experienced during development.

Earlier influences from the environment can also affect the organism by interfering with the acquirement of new experiences. Ideally, man should remain receptive to new stimuli and new situations in order to continue to develop. In practice, however, the ability to perceive the external world with freshness decreases as the senses and the mind are increasingly conditioned in the course of life.

Complete receptivity is the prerogative of childhood and of the few privileged adults who have retained or recaptured the directness of perception which enables most children to see "things as they are." Hence the deep biological truth of Baudelaire's arresting image, "*Le génie, c'est l'enfance retrouvée*" (Genius is childhood recaptured).

Recent studies suggest that the effects of imprinting and other early influences may not be as irreversible as is generally believed and can be erased by various psychological manipulations and perhaps by drugs.[27] The ancient dream of the Fountain of Youth might acquire a new and richer meaning if acceptable techniques could be developed to reestablish a state of receptivity in the fully developed adult. Recapturing childhood, in Baudelaire's sense, could then mean reacquiring the ability to perceive the external world directly.

To sum up, it seems legitimate to assume that all changes in ways of life—not only the spread of technology and of scientific knowledge—continuously alter the perceptual world of the developing organism. New behavioral patterns and new problems of social adaptation inevitably result from such environmental changes; these in turn impart to individuality some characteristics that are shared by most members of a given generation.

In the final analysis, individuality emerges progressively from the manner in which each person turns all experiences of the body and the mind into a knowledge so structured that it can be used for further growth and for action.

Of Human Freedom

The biological sciences deliberately attempt to account for man's nature without reference to free will; they concern themselves with the deterministic aspects of life. Determinism, as used here, denotes complete predictability, given the momentary conditions, the pertinent laws, and the proper mathematical and other techniques required to predict consequences by integrating the relevant information. This approach has provided much knowledge concerning the natural history of man—his origins, his evolution, his biological characteristics, the mechanisms of his behavior, and the determinants of his social structures, but it has thrown no light on the nature of human freedom.

Whether they deal with plants or animals, microbes or men, biologists never have any difficulty in demonstrating that the inherited genetic constitution governs the development of all anatomical structures, physiological characteristics, and mental processes. They can also establish that various forms of experience and learning have effects which persist for the whole life span and that environmental stimuli elicit certain predictable responses from the organism at every moment of its existence. Since all biological and mental processes are conditioned by the genetic endowment, past experiences, and environmental factors, one might conclude that there is no place in life for individual freedom. However, the failure of biologists to recognize in their studies the manifestations of freedom may have its origin in the methods they use. The very nature of the experimental method leads scientists to focus their efforts on phenomena that are reproducible and therefore are largely independent of free will.

Even though all manifestations of life are known to be conditioned by heredity, past experiences, and environmental factors, we also know that free will enables human beings to transcend the constraints of biological determinism. The ability to choose among ideas and possible courses of action may be the most important of all human attributes; it has probably been and still is a crucial determinant of human evolution. The most damning statement that can be made about the sciences of life as presently practiced is that they deliberately ignore the most important phenomena of human life.

Many attempts have been made by philosophers and

scientists to find ways of reconciling determinism and free will. It is not surprising that these attempts have been largely unsuccessful; the history of science shows that complex phenomena can rarely be explained in terms of simpler and more restricted phenomena. Light cannot be understood by regarding it merely as a stream of particles moving in accordance with the laws of classical mechanics. The understanding of electromagnetism required the concept of fields, which was revolutionary at the time it was introduced. Cosmology and certain types of submicroscopic processes achieved a new depth of understanding with the development of the general theory of relativity and of quantum mechanics.[28]

These examples, among many others that could be cited, suggest that the understanding of consciousness, free will, and other truly human manifestations of life will also require new concepts different from and complementary to present biological theories. The attitudes toward the problems of life identified with the expressions free will and determinism may come to appear less incompatible if living processes are some day shown to involve concepts more subtle than those that dominate contemporary physics, chemistry, and biology.

Investigators concerned with the mental processes involved in choices and decision-making have pointed out, furthermore, that it may not be fundamentally and logically possible for the human brain to achieve a complete understanding of itself and of its workings, any more than a computing machine can completely predict the future of a universe of which it is a part.[29]

Niels Bohr saw in the determinism-freedom polarity a manifestation of the complementary principle that he had formulated to account for the fact that an electron under certain conditions behaves like a wave, and under others like a particle. Yet when the process of decision-making is analyzed in detail, and each step followed in its causal connections, freedom seems to disappear, because all aspects of behavior are affected by genetic, experiential, and environmental factors. Just as physicists study the electron either as a wave or as a particle depending on the conditions under which its behavior is observed, so, Bohr asserted, human behavior can probably be studied as expressions of free will or determinism depending upon the point of view of the observer.[30]

In any case, the limits and potentialities of freedom have deterministic components that reside in environmental forces and in the innate and acquired biological characteristics of each individual person. While every human being is unprecedented, unique, and unrepeatable, by virtue of his genetic constitution and past experiences, his environment determines at any given moment which of his physical and mental potentialities are realized in his life. Free will can engender acts of freedom only to the extent that past and present conditions make it possible to actualize concepts and anticipations. An example is the fact previously mentioned that slum children acquire early in life a culture from which they find it difficult to escape; their surroundings destroy much of their potential freedom.

The English geneticist J. B. S. Haldane (1892–1964) expressed with precision the extent to which the possibility of exercising freedom is conditioned by the total environment. "That society enjoys the greatest amount of liberty," he wrote, "in which the greatest number of human genotypes can develop their peculiar abilities. It is generally admitted that liberty demands equality of opportunity. It is not equally realized that it demands a variety of opportunities."

Since the physical and social environment plays such a large role in the exercise of freedom, environments should be designed to provide conditions for enlarging as much as possible the range of choices. This applies to social planning, urban or rural development, and all the practices that affect the conduct of life. In most situations, design could certainly be improved by a better knowledge of man's nature and of the effects that the environment exerts on his physical and mental being. But design involves also matters of values, because free will can operate only where there is first some form of conviction.

Values are sometimes considered to be unchangeable because they are believed to be built into man's innate moral nature. In practice, however, many of the values by which men operate are based on prevailing social attitudes, as well as on inclinations, prejudices, and the common sense derived from the experience of daily life. There is also a real possibility that, in the future, values might increasingly originate from the natural and social sciences. Scientific knowledge *per se* cannot define or impose values to govern behavior, but it provides facts on the basis of which choices can be made.

While choice can be made more rational by basing it on factual information, and on evaluation of consequences, it always retains a personal component because it must ultimately involve a value judgment. This constitutes another expression of the determinism-freedom polarity, which is one of the most characteristic aspects of the human condition.

Freedom is concerned not only with what to do, but perhaps even more with what not to do. Angels are not free,

Saint Augustine reminds us, because they are not able to sin. Man is free because he is able not to sin; he has a choice. A painting of a landscape can be superior to a photograph of it, because the painter has greater freedom than the photographer in leaving out of the composition the parts not relevant to the atmosphere or message he wants to convey.

The British scientist and philosopher Jacob Bronowski has stated that "the explosive charge which, in this century, has split open the self-assurance of Western man" is contained in "the bland proposition that man is a part of nature." If it is true, as it seems to be true, Bronowski wrote, that "living matter is not different in kind from dead matter," then "man as a species will be shown to be no more than a machinery of atoms" and if this is so, he cannot be a "person."[31]

To describe man as "no more than a machinery of atoms" provides but a very incomplete account of his nature. The methods used by the investigator determine and limit the kind of observations he can make. If scientists elect to study man only by physicochemical methods, they will naturally discover only the physicochemical determinants of his life and find that his body is a machinery of atoms. But they will overlook other human characteristics that are at least as interesting and important. One of them is that man hardly ever reacts passively to external forces. The most characteristic aspect of his behavior is that he responds not only actively but often unexpectedly and creatively. He is the more human the more vigorously he converts passive reactions into creative responses. The mechanical definition of human life misses the point because what is human in man is precisely that which is not mechanical.

In fact, it is not likely that the orthodox mechanical

definition of life applies to animals either. According to the Harvard biologist George Wald, the unpredictability of animal behavior led an exasperated physiologist to state what has come to be known as the Harvard Law of Animal Behavior: "Under precisely controlled conditions, an animal does as he damn pleases." In Wald's words, "Could one ask more free will than that?"[32]

That the same view is probably held in biological laboratories all over the world is indicated in a recent autobiographical article by the Belgian physiologist and Nobel Prize laureate Corneille Heymans: "In our laboratory, a large picture, a cartoon of a funny dog, is hanging on the wall. The dog is looking at a syringe ready to give him an injection. Under the dog, the statement of Nickerson is written: 'Under the most perfect laboratory conditions and the most carefully planned and controlled experimental procedures, animals will do what they damned please!' "[33]

Human freedom includes the power to express innate potentialities, the ability to select among different options, and the willingness to accept responsibilities. All these and other such forms of activity involving choice and volition transcend the kind of determinism that would account for the operations of a machine. The very use of the word "machine" in fact points to the conceptual difficulties presented by a narrow view of determinism. To consider man or any living thing as a machine implies the assumption that it works to some designated end. Even a part of a machine serves its particular end. Man's body is a machinery of atoms in the sense that its structures and functions obey the laws of inanimate matter. But it is a machine also in the sense that man himself can control its operations toward certain goals—the life goals that he freely selects. Even among the most orthodox materialists there are few who would not agree with the religious philosopher Paul Tillich (1886–1965) that man becomes really human only at the time of decision, when he exercises free will.

Man's most characteristic attribute, his conscious orientation toward the future, implies willingness to make the efforts required for shaping his personality and thereby his destiny. In the words of José Ortega y Gasset (1883–1955): "Living is precisely the inexorable necessity to make oneself determinate, *to enter into an exclusive destiny*, to accept it—that is, to resolve to *be it*. We *have*, whether we like it or not, to realize our 'personage,' our vocation, our vital program, our 'entelechy'—there is no lack of names for the terrible reality which is our authentic I (ego)."[34]

In *Notes From Underground*, Feodor Dostoevski's sniveling hero could not find satisfaction in the order and comfort of the "Crystal Palace" world in which he lived; he chose an antisocial way of life because it was the one form of freedom of action still available to him. "I say, gentlemen, hadn't we better kick over the whole show and scatter rationalism to the winds, simply to send these logarithms to the devil, and to enable us to live once more at our own sweet foolish will? . . . the whole work of man really seems to consist in nothing but proving to himself every minute that he is a man and not a piano-key."[35] Dostoevski's man will affirm his individuality even if it means physical suffering and even if it means turning his back on civilization.

In a similar vein Tillich has repeatedly defended the view that individualism is the self-affirmation of the individual self without regard to participation in its world. These words are dangerous if interpreted to mean that freedom is an absolute value without relevance to social relationships.

Human freedom does not imply anarchy and complete permissiveness. Such attitudes would inevitably result in the disintegration of individual lives and of the social order. Rejection of discipline is unbiological because it is incompatible with physical, mental, and social health, indeed with the survival of the human species. Design, rather than anarchy, characterizes life. In human life, design implies the acceptance and even the deliberate choice of certain constraints which are deterministic to the extent that they incorporate the influences of the past and of the environment. But design is also the expression of free will because it always involves value judgments and anticipates the future.

5. THE PURSUIT OF SIGNIFICANCE

Surroundings and Events

Western man tends to consider himself apart from and above the rest of creation. He has accepted to the letter the Biblical teaching that man was given by God "dominion over the fish of the sea, and over the fowl of the air, and over the cattle, and over all the earth" (Genesis 1:26).

In contrast, most primitive people identify themselves with the environment in which they are born and live. They worship the sky and the clouds, trees and animals, mountains, rocks, springs, and rivers as the living expressions of the cosmic order from which they derive their own being. Man's feeling of identity with Nature was beautifully expressed by the Indian chief Seattle in an address to Governor Isaac Stevens, Commissioner of Indian Affairs for the Territory of Washington. The occasion was a ceremony in 1853 or 1854 during which the Governor presented to the Indians the terms of a treaty for the surrender of the land on which the city of Seattle is now located. A few years later Dr. Henry A. Smith, who had witnessed the ceremony, reported it in the following words:

"Old Chief Seattle was the largest Indian I ever saw, and by far the noblest-looking. He stood nearly six feet in his moccasins, and was broad-shouldered, deep-chested, and finely proportioned. His eyes were large, intelligent, expressive and friendly when in repose, and faithfully mirrored the varying moods of the great soul that looked through them. . . .

"When rising to speak in council or tendering advice, all eyes were turned upon him, and deep-toned, sonorous and eloquent sentences rolled from his lips like the ceaseless thunder of cataracts."[1]

After being presented with the text of the settlement, Chief Seattle placed one hand upon General Stevens' head, slowly pointed the index finger of the other hand heavenward, and solemnly made his reply. His words were later translated by Dr. Smith as follows:

"There was a time when our people covered the whole land as the waves of a wind-ruffled sea covers its shell-paved floor, but that time has long since passed away with the greatness of tribes now almost forgotten. I will not dwell on nor mourn over our untimely decay, nor reproach my paleface brothers with hastening it, for we, too, may have been somewhat to blame. . . .

"We are two distinct races, and must ever remain so, with separate origins and separate destinies. There is little in common between us.

"To us the ashes of our ancestors are sacred and their final resting place is hallowed ground, while you wander far from the graves of your ancestors and, seemingly, without regret. . . .

"Our dead never forget this beautiful world that gave them being. They still love its winding rivers, its great mountains, and its sequestered vales. . . .

"Every part of this country is sacred to my people. Every hillside, every valley, every plain and grove has been hallowed by some fond memory or some sad experience of my tribe. Even the rocks, which seem to lie dumb as they swelter in the sun along the silent seashore in solemn grandeur, thrill with memories of past events connected with the lives of my people.

"The very dust under your feet responds more lovingly to our footsteps than to yours, because it is the ashes of our ancestors, and our bare feet are conscious of the sympathetic touch, for the soil is rich with the life of our kindred"[2] [italics mine—R.D.].

Indians of other tribes have similarly expressed a feeling of organic unity with their ancestral lands. A few decades ago, the Navajos, protesting against new federal regulations which limited (for very good reasons) their grazing practices, reminded the government officials that ". . . before we were born, the white people and our old folks made a treaty. The treaty was made to the end that these encircling Mountains would always be ours, so that we could live according to them. The right to these was given to us, so that all the Navajos might live in accord with that which is called Mountain Soil, and the pollen of all plants. All Navajos live in accord with them."[3]

The Southwest is so different from the Northwest in climate, topography, and natural resources that there is little in common between the ways of life and traditions of the Navajos and those of Chief Seattle's tribe. But both tribes shared a mystic sense of relationship with the natural forces of their ancestral surroundings. Their lives derived significance from an emotional identification with Nature.

Before the industrial age, people had everywhere achieved some measure of integration with the physical and social environment in which they lived and on which they depended. When he first reached Europe from India, the philosopher Sir Rabindranath Tagore (1861–1941) marveled at the extent to which the quality of the European countryside was a loving creation of the peasantry, the result of an active wooing of the earth.[4] One of the reasons for the emotional impoverishment in countries where industry and technology have taken over is the loss of identification with the natural world. Increasingly we tend to deal with nature as if it were of value only as a source of raw material and entertainment.

Chief Seattle had good reasons for believing that the spirit of Western civilization was antithetical to the Red Man's ways of life and feeling for Nature. He was also right in stating to Governor Stevens, still according to Dr. Smith's account:

"Your religion was written on tablets of stone by the iron finger of an angry God. . . .

"Our religion is the traditions of our ancestors—the dreams of our old men given to them in the solemn hours of night by the Great Spirit . . . and is written in the hearts of our people."[5]

However, the great difference in attitude of the red man and the white man was not caused by their belonging to "two distinct races . . . with separate origins and separate destinies." Whether man considers himself part of nature, or outside of it and its master, is determined not by racial origins, but by cultural forces. Racial differences have little relevance to mental or emotional characteristics, and do not necessarily imply separate destinies. Within the past few decades, many Indians whose racial purity has not been diluted by mixed marriage have nevertheless become completely westernized and thereby lost the sense of identification with their ancestral lands and ways of life. In fact, this is particularly true of the Indians of the Northwest, in whose name Chief Seattle made his memorable address. Likewise, many Europeans who have migrated to the Americas, North and South, have now renounced their allegiance to the lands and civilizations of their origin.

All men are migrants from a common origin. For all of them life can be sustained only within extremely narrow physical limits defined by the physiological exigencies of the species *Homo sapiens.* But men of all races can learn to live in a wide variety of natural environments by adopting the proper ways of life. While man's physiologic adaptations are rather limited in range, he has learned to supplement them with sociocultural adaptations which are becoming increasingly effective and diversified.

Whatever their races and social origins, human beings have established their abodes under every possible type of physical condition; they have also developed emotional ties to all manifestations of nature on earth. *Homo sapiens* probably originated in a temperate climate, but human beings of all racial types have now made themselves at home under the sun of the Sahara and the fogs of Newfoundland, in the low-lands of the African rain forest and in the high Peruvian Andes. Human social life started in small isolated bands, but it is now happily carried out in huge crowded cities as well as in cozy villages. Home is that environment to which a particular person becomes adapted through biological and sociocultural mechanisms, and to which he becomes emotionally attached through the traditions of his group and his own personal experiences. Home is less a physical place than a locus with which past experiences are identified.

Throughout history, in all parts of the world, populations have been compelled to abandon their homes and to resettle in other lands, as a result either of wars or of natural disasters, or for economic or ideological reasons. But human adaptability is so great that displaced populations have usually succeeded in re-creating a home with all the connotations of the word even when the move had taken them to entirely different physical and human surroundings. Although most of the highland populations in Britain are Celtic in descent, they did not originate in the areas to which they now seem so well adapted; they were natives of the lowlands and were driven out by Saxon invaders.[6] Similarly, the northeast coast of the United States has become within a few generations the home of Irish, Italians, Jews, and Central Africans who were forced out of their original homelands for a variety of economic and social causes.

Daily life in all the large cities of Western civilization provides endless examples of man's ability to function in physical and social surroundings totally different from those of his origins. Most human beings are potentially able to integrate their biological and social past with that of their neighbors and contemporaries. In the Old World as far north as London, and in the New World as far south as Buenos Aires, one can see—working at the same tasks, playing at the same games, acquiring the same habits—blond blue-eyed Scandinavians or Celts born in misty lands; Arabs and Indians originating from sandy deserts; dark-skinned Africans or yellow-skinned Asians whose parents lived in tropical rain forests. All these people eventually eat the same food, listen to the same music, watch the same spectacles, pledge allegiance to the same flag, worship the same God, or become equally indifferent to any form of traditional religion. They tend to forget their ancestral heritages, commonly suffer in much the same way from loss of traditional values, and together clumsily search for a new significance in life.

Irrespective of race and color, most human beings can also develop tolerance to a large variety of conditions that are certainly undesirable and may indeed appear at first sight almost unbearable. The Office of Civilian Defense recently made an exhaustive review of what is known concerning the effects on health of extreme physical crowding, such as occurred, for example, in slave ships, concentration camps, prisons, and bomb shelters.[7] Many people survived these horrible experiences and recovered rapidly from them because they controlled tension and reserved their energy during the ordeal; they allowed their systems to adjust by submitting to the stresses almost passively!

Two contemporary examples also illustrate the surprising range of conditions which human beings can learn to tolerate and even to enjoy. At the Valley Forge Interchange in Pennsylvania there is emerging a huge development accommodating 25,000 industrial employees, 1 million square feet of retail space, 1,000 acres of parking lots, expressways, and thoroughfares—but not a single family dwelling! Another instance of modern man's ability and willingness to let his life be mechanized for the sake of professional activities is the fact that the Los Angeles airport is becoming the largest office and hotel district on the West Coast. Many persons regard such a peculiar environment as a most satisfactory place for business and other meetings. How remote seem the palatial mansions in which government officials and business tycoons used to gather for state and business transactions!

History shows that cultures of a sort can emerge from the most improbable ways of life, provided these last long enough to become integrated into an organic whole. The emergence of a new culture is rarely if ever the result of a conscious choice with a definite goal in mind. What happens rather is that the social customs for mating and raising children, providing shelter and means of subsistence, developing natural resources, protecting the land against enemies, enjoying life, or worshiping God interact and become organized into unique patterns. Societies of all types, from the simplest to the most complex, have achieved such integration of behavioral activities and thus have given rise to the marvelous diversity of human cultures.

The Australian aborigines, primitive as they are, have developed extremely sophisticated customs that enable them to survive under the harsh conditions of the bush.[8] These customs can hardly be traced to a systematic and entirely conscious exercise of human intelligence; they must have emerged progressively in the course of continued interactions between the people and their environment. The aborigines' responses to the conditions of their lives and the tribal memories of their past experiences have become incorporated in their social wisdom. Similarly, it is most improbable that any group of Englishmen, however intelligent and foresighted, could have devised in the abstract and then imposed on their fellowmen a scheme as complex as that of the British type of parliamentary government. Its formulation was the product of many successive pragmatic adjustments rather than of a complete plan thought out in its entirety and in all its details.

Biologically, man is still the great amateur of the animal kingdom; he is unique in his lack of anatomical and physiological specialization. The range of his adaptive potentialities has been greatly enlarged by sociocultural mechanisms that have enabled him to colonize most of the earth. His adventurous spirit now tempts him to conquer other worlds. But despite the success of launchings into space, his colonizing days are over.

Science-fiction writers and a few scientists notwithstanding, man will never be able actually to settle anywhere in the cosmos other than on or near the surface of the earth. At most, he will make hit-and-run raids on the moon, Mars, and perhaps other planets; he may also establish some stations for specialized purposes under the ocean waters. But he is earthbound forever because his life is completely dependent on fresh water and especially on the earth's atmosphere. While it is possible to re-create and maintain the earth's atmosphere out in space or on the bottom of the ocean, this technological enterprise is so formidable that it will be done only for very special missions.

The fact that modern man is constantly moving into new environments gives the impression that he is enlarging the range of his biological adaptabilities and thus escaping from the bondage of his past. This is only an appearance. Wherever he goes, and whatever he does, man is successful only to the extent that he functions in a microenvironment not drastically different from the one under which he evolved. He can climb Mount Everest and fly at high altitudes only if he carries an adequate oxygen supply and is equipped to protect himself against cold. He moves in outer space and at the bottom of the sea only if he remains within enclosures that almost duplicate the terrestrial environment or links himself to the earth by an umbilical cord. Even the Eskimos, who appear so well adapted to the Arctic climate, in reality cannot long resist intense cold. Sheltered in their snow houses or clothed in their parkas, they live an almost tropical life![9]

There is no hope whatever that man's biological nature can be changed enough to enable him to survive without the earth's atmosphere; in fact, the very statement of this possibility is meaningless. *Homo sapiens* achieved his characteristics as a biological species more than 100,000 years ago, and his fundamental biological characteristics could not be drastically altered without destroying his very being. He developed his human attributes in the very act of responding to the environment in which he evolved. The earth has been his cradle and will remain his home.

The experiences of the present century show that mankind has not lost the biological adaptability that enabled *Homo sapiens* to become established over most of the earth. Countless human beings have survived the frightful ordeal of combat during war, and many managed to function even in the worst concentration camps. In our own communities today, human life is adjusting itself to the multifarious physiological and mental stresses of urban and industrial environments. All over the world, indeed, the most polluted, crowded, and brutal cities are also the ones that have the greatest appeal. Some of the most spectacular increases in population are occurring in areas where living conditions are detestable from all points of view.

Modern man can adjust to environmental pollution, intense crowding, deficient or excessive diet, as well as to monotonous and ugly surroundings. Furthermore, biologically undesirable conditions do not necessarily constitute a handicap for economic growth. Great wealth is being produced by men working under extreme nervous tension amidst the infernal noise of high-power equipment, telephones, and typewriters, in atmospheres contaminated with chemical fumes or in crowded offices clouded with tobacco smoke.

Such adaptability is obviously an asset for survival and seems to assure the continued biological success of the human race. Paradoxically, however, the very fact that man readily achieves biological and sociocultural adjustments to so many different kinds of stresses and undesirable conditions is dangerous for his welfare and his future. But before justifying this statement, I must point out that the classical meanings of the term adaptation do not properly apply to the adjustments that human beings have to make under the conditions of modern life.[10]

For the general biologist, Darwinian adaptation implies a state of fitness to a given environment, enabling the species to survive and multiply. In this light, man is remarkably adapted to life in highly urbanized and industrialized societies; his populations continuously increase and he spreads urbanization and industrialization over more and more of the earth. Even if it is true that the modern ways of life do not contribute significantly to real happiness and may even increase the frequency of certain chronic disorders, these failures are of little importance from the purely biological point of view. The chronic disorders characteristic of modern civilization affect man chiefly during late adulthood after he has fulfilled his reproductive functions and contributed his share to social and economic development. The problem of happiness becomes important only when attention is shifted from the purely biological aspects of life to the far different problems of human values. In applying to man the concept of adaptation, we must therefore use criteria different from those used in general biology.

Physiologists or psychologists give the word adaptation a broader meaning than that associated with Darwinian population theory, but they, too, seem to underestimate the rich complexity of human life. To them, a response is adaptive when it promotes homeostasis—in other words, when it brings into action the metabolic, hormonal, or mental processes that tend to correct the disturbing effects of environmental forces on the body and mind. Such adaptive responses obviously contribute to the welfare of the organism at the time they occur, but unfortunately they commonly have secondary effects that may become deleterious.

Scar tissue heals wounds and helps in checking the spread of infection; it represents a successful homeostatic process at the time it is formed in response to a wound or a lesion. But scar tissue in the liver or in the kidney is responsible for serious diseases such as cirrhosis or glomerular nephritis; in the lungs it may seriously impede breathing; in the joints it generates the frozen immobility of rheumatoid arthritis. Similarly, many other medical problems have their origin in biological and mental adaptive responses that allowed man to cope with environmental threats earlier in life. All too often the wisdom of the body is a short-sighted wisdom.[11]

Man's physiological and psychological endowments thus give him a wide range of adaptive potentialities and enable him to survive and function even under extremely unfavorable conditions; however, the fact that all subsequent aspects of his life are affected by his past makes such adaptability a double-edged sword. Evaluated over the entire life span, the homeostatic mechanisms through which adaptation is achieved often fail in the long run because they result in delayed pathological effects.

Countless incidents in the life of urban-dwellers illustrate how we overlook the dangers that result from our temporary adjustments to undesirable situations.

On a hot and humid Friday during midsummer, I landed at Kennedy Airport early in the afternoon. The taxicab that was taking me home was soon caught in a traffic jam, which gave the driver an opportunity to express his views on the state of the world. Noting my foreign accent, he assumed that I was unacquainted with the United States and proceeded to enlighten me on the superiorities of American life. "You probably are surprised by this heavy traffic so early on Friday afternoon," he remarked, as the cab stood still in the sultry air saturated with gasoline fumes. "The reason there are so many people on the road at this hour is that we have plenty of leisure in this country and all of us can afford an automobile." As we removed our coats and mopped our brows, he added forcefully, "In the United States we all live like kings."

Since I was irritated by the delay, the driver's statement that we lived like kings appeared to me completely irrational, as if coming from a deranged person. But he looked like a reasonable man, similar in outward appearance and behavior to the thousands of other automobile occupants who were spending the afternoon on the congested and ill-smelling road still called by poetic license an expressway.

Like millions of other persons breathing gasoline fumes and struggling with crowds all over the country on that midsummer Friday afternoon, my taxi driver had made some sort of adjustment to air pollution, to competition with countless anonymous motorists, and to the dismal monotony and ugliness of the suburban scenery. In fact, he was so well adjusted that he could carry out his professional activities with great effectiveness and thus contribute to the growth of the national economy. The pollution and other stresses that he experienced daily on the crowded highways increased the likelihood that he would eventually suffer from some physical and behavioral disturbances. But degenerative disorders would become manifest only in his late adulthood, after he had produced a family and played his role in the economic enterprise. He would not trace them to the nonsensical ways of life that he had come to accept as part of living like a king.

For some two centuries, Western man has believed that he would find his salvation in technology. Unquestionably technological innovations have increased his economic wealth and improved his physical health—although they have not necessarily brought him the kinds of wealth and health that generate happiness. Technology provided my taxi driver with all the raw materials required for building a large body and a reasonably equipped mind but it also imposed on him ways of life almost incompatible with the maintenance of physical and mental sanity. In this regard, his plight was symbolic of life in highly technicized societies. The precise causes of the diseases of civilization are difficult to identify, but there is no doubt that many originate directly or indirectly from deleterious environmental influences to which human beings *seem* to become adjusted.

Life expectancy at birth has increased enormously in all industrialized countries and is now 70 years or more in the prosperous social groups. However, life expectancy past the age of 45 has not significantly increased anywhere in the world, and especially *not* among the social groups that enjoy great comfort and can afford elaborate medical care; it may even be somewhat lower in very prosperous countries such as the United States or Sweden than in economically less-favored countries such as Spain or Ireland.[12]

The control of mortality during the early years of life accounts for the increase in life expectancy at birth all over the world, but the failure to control vascular disorders, malignancies, and other degenerative diseases has prevented so far any significant increase in true longevity. Relative to the total adult population, the percentage of nonagenarians and centenarians is probably no greater today than it was in the past. Adults have to pay in the form of chronic and degenerative diseases for the unrecognized insults they have received from the environment in their earlier years.

Many of the health problems of modern man, in the present and for the future, have their origin in slowly developing injurious effects of the technological environment and the new ways of life. It has long been known for example that radiation damage may not become apparent for one or two decades; the consequences of cigarette smoking and of environmental pollution likewise develop very slowly; certain medical and psychic problems of adult life have their origin in malnutrition and other forms of deprivation experienced during infancy. The biological remembrance of the past is thus of particular relevance to the understanding of the diseases that affect adult life and old age.

Atmospheric pollution in the industrial areas of northern Europe provides striking examples both of man's ability to function in a biologically undesirable environment and of the remote dangers inherent in this adaptability. Ever since the beginning of the Industrial Revolution, the inhabitants of northern Europe have been heavily exposed to many types of air pollutants, some produced by incomplete combustion of coal and others released in the fumes from chemical plants; these pollutants are rendered even more objectionable by the inclemency of the Atlantic climate. However, through long experience with pollution and with bad weather a variety of adaptive physiological reactions and living habits have progressively developed. Northern Europeans accept their dismal environment almost cheerfully, even though such conditions appear unbearable to outsiders who experience them for the first time.

Adaptive responses to environmental pollution occur in heavily industrialized areas all over the world. Since people can function effectively despite the almost constant presence of irritating substances in the air they breathe, one might assume that human beings can make adequate adjustments to massive air pollution. These adjustments are inadequate in the long run, however, because air pollution eventually causes suffering and economic loss. Even among persons who are almost unaware of the air pollutants surrounding them, the respiratory tract continuously registers the insult. After periods of time that differ from one case to another and commonly extend to several decades, the cumulative effects of irritation become manifest in the form of chronic bronchitis and other types of pulmonary disease. Since these pathological consequences do not occur until long after initial exposure, it is always difficult to relate them to the primary physiological insult, which may have been so mild as to have remained unnoticed.

Chronic pulmonary disease now constitutes the greatest single medical problem in northern Europe, as well as the most costly. It is increasing in prevalence at an alarming rate also in North America and will spread to all areas undergoing industrialization. Furthermore, air pollution probably increases the prevalence of various types of cancers and of fatalities among persons suffering from vascular diseases. Here again, the long time span between cause and effect makes it difficult to establish the causal relationships convincingly.

The delayed effects of air pollutants constitute models for the kind of medical problems likely to arise in the future from other forms of environmental pollution, such as the pollution of water and food by products of industry. Granted differences in detail, the course of events can be predicted in its general trends.

Wherever convenient, chemical pollution of air, water, and food will be sufficiently controlled to prevent immediately disabling or obviously unpleasant toxic effects. Human beings will then tolerate, without complaint, concentrations of pollutants that do not interfere seriously with social and economic life. This continued exposure to low levels of toxic agents will eventually result in a great variety of delayed pathological manifestations, creating physiological misery, increasing the medical burden, and lowering the quality of life. The point of importance here is that the most significant effects of environmental pollutants will not be detected at the time of exposure to them; indeed, they may not become evident until several decades later. In other words, society will become adjusted to levels of pollution sufficiently low not to have an immediate nuisance value, but this adjustment will eventually interfere with the enjoyment of later life and also with the full expression of its potentialities. Malnutrition or overnutrition, minor infections, and the indiscriminate use of drugs are other aspects of life which are likely to have long-range consequences far more dangerous than their immediate effects.

In view of the increase in the world population, the problems posed by adaptation to crowding will certainly change in character and become more important in the near future. Man is a gregarious animal; he generally tends to accept crowded environments and even to seek them. While this attitude unquestionably has social advantages, these may not be unmixed blessings. Physiological tests have revealed that crowding commonly results in an increased secretion of various hormones which affect the whole human physiology. An adequate hormonal activity is essential for well-being but any excess has a variety of harmful effects.

As the world becomes more and more urbanized and industrialized, constant and intimate contact with hordes of human beings has come to constitute the "normal" way of life, and men have eagerly adjusted to it. This change has certainly brought about all kinds of behavioral adaptations to social environments that appear normal to us even though they would have been shocking and possibly disastrous in the past. Crowding is a relative term; the past experience of the group conditions the manner in which each of its members interacts. Population density, in other words, is probably less important in the long run than the intensity of social conflicts it brings about, conflicts which usually become less intense after social adjustments have been made. Granted that the consequences of crowding are not yet well understood, there is little doubt that they will be found in most cases to have an insidious course. The worst effects will not be the initial ones, but the complex secondary responses called forth later in individual persons and in society as a whole.

During evolutionary development the human species probably functioned in small groups, each member knowing the others personally. Man may have need for larger gatherings now and then, but certainly not as a constant diet. When oversocialized, he is likely to react with frustrations, repressions, or aggressions that may evolve into neuroses.

Admittedly, children can be reared and trained in an environment so oversocialized that they no longer feel happy or safe outside a crowd of their own kind. This situation only illustrates once more that human beings can become habituated to conditions that are undesirable in the long run. Like adults, children can be habituated to search for happiness in overeating, unbalanced food, unsuitable amusements, perverted addictions. Such habituations provide temporary relief or even satisfaction, but they are of course dangerous. This is probably true of habituation to overcrowding. Architects have shown that more ingenious design of human settlements can compensate to some extent for insufficient space, but there are limits to what can be achieved by architectural ingenuity.[13] Beyond these limits, overcrowding is likely to cause psychological damage. To some overcrowded populations violence or even the bomb may one day no longer seem a threat but rather become a release.

Complete surrender to overcrowding in a highly technicized society is not likely to destroy mankind, but it will mean an increasingly organized world. The environment will favor the selective reproduction of people best suited to a regimented life. Many people today are maladjusted to crowded life, but as long as there are uncrowded places and social control remains ineffective, those who really want to enjoy a free life can still find a world of their own choosing. If crowding and regimentation continue to increase, however, the descendants of such maladjusted people will be progressively eliminated. When they disappear, many of our present human values will become meaningless and will eventually be forgotten. There will be no place for sensitive literature, intensely personal art, or unorthodox science in the human ant hill of the future; not even room for primitive Christianity. What meaning can the parables and poetry of the past retain if there are no lilies in the field? We must hope that there will still be rebels to champion freedom.

High levels of prosperity can thus create a whole range of undesirable situations accumulating throughout the whole life span. Environmental pollution, excessive food intake, lack of physical exercise, the constant bombardment of stimuli, the inescapable estrangement of civilized life from the natural biological rhythms are but a few among the many manifestations of these. An inevitable result of urbanization, of population growth, and, paradoxically, of higher living standards is that the affluent society is also, as some wags have called it, the "effluent" society.

It should be possible to identify the environmental factors responsible for the chronic and degenerative disorders of our societies. But even when this has been done, it may prove extremely difficult to control these factors because all aspects of the urban and industrial environment are intimately interwoven in the social fabric. Furthermore, the biological and social environment can hardly be dissociated from individual behavior.

Keeping streets and houses clear of refuse, filtering and chlorinating the water supplies, watching over the purity of food products, assuring a safe minimum of fresh air in public places constitute measures that can be applied by the collectivity without interfering seriously with individual freedom. These measures are readily accepted because they do not demand personal effort from their beneficiaries. In contrast, any measure that requires individual discipline is more likely to be neglected. Almost everybody is aware of the dangers associated with overeating, lack of physical exercise, chain-cigarette smoking, excessive consumption of alcohol or drugs, or exposure to polluted environments. But few are the persons willing to make the individual efforts required to avoid these dangers. Furthermore, the consequences of environmental threats are so often indirect and delayed that the public is hardly aware of them. Many effects of the environment become inscribed in the body and the mind without the affected person's realizing that he is being changed irreversibly by influences that do not enter his consciousness.

The adjustments to environmental threats mentioned in the preceding paragraphs are relevant to the pursuit of life's higher significance for several reasons. One is simply that the state of health conditions human response to any situation. Another reason is that adjustments to environmental threats are often achieved through a blunting of awareness and can thereby interfere with the recognition of human values. Most persons come to be almost unaware of conditions which they know to be undesirable but to which they have become tolerant through continued exposure.

Smogs, unpleasant odors, and other forms of environmental pollution, noise from street and air traffic, crowding and excessive stimuli are but a few among the common manifestations of modern life that are extremely objectionable when first experienced, then progressively escape conscious awareness. Few are the urban dwellers, even among the sensitive, who realize that they hardly ever experience fragrant air or a starry night. Most of us become oblivious to the filth, visual confusion, dirt, and outright ugliness that we encounter morning and night on our way to and from the office.

Similarly, when changes occur progressively in social life, the quality of human contacts can degenerate without the persons involved being conscious of the loss this entails. As one's social world enlarges, the number of acquaintances increases, but the depth of relationships usually decreases. Urban dwellers may have many friends, but the nature of the friendship is commonly superficial and rarely corresponds to the quality of Martin Buber's "I-Thou" encounter. In its present form, urban life makes it difficult to maintain the intimate face-to-face association and cooperation experienced in the small groups of people that some psychologists designate as "primary groups." The widespread nostalgia for school or army days probably constitutes in many cases an unconscious acknowledgment of emotional hunger. Ordinary adult life rarely provides the enriching experience of close comradeship and mutual dependence which was possible when youthful generosity and constant associations gave a chance for really meaningful social encounters.

One of the worst consequences of modern life, according to the American urban planner Christopher Alexander, is the "autonomy-withdrawal" syndrome. Most people, he claims, use their homes to escape from the stresses of the outside world and practice social withdrawal as a form of self-protection. Eventually withdrawal becomes a habit; people reach a point where they become unable or unwilling to let others penetrate their own private world.[14]

Extreme individualism and autonomy commonly develop unconsciously as a consequence of the self-protective withdrawal from stress. Persons who have achieved such autonomy remain dependent on the social groups of which they are a part, but this dependence manifests itself almost exclusively through the medium of money. Money in turn tends to create a world reinforcing individualism and withdrawal. According to Alexander, the fact that the song "People who need people are the luckiest people in the world" made the top of the United States hit parade in 1964 indicates that this pathological individualism is very widespread. Unfortunately, few people are aware of the impoverishment of life resulting from the autonomy-withdrawal syndrome; even fewer seem to realize that the pursuit of significance is bound to fail unless man learns once more to speak to man.

History confirms present-day observations in demonstrating that man can become adjusted, socially and biologically, to ways of life and environments that have hardly anything in common with those in which civilization emerged and evolved. He can survive, multiply, and create material wealth in an overcrowded, monotonous, and completely polluted environment, provided he surrenders his individual rights, accepts certain forms of physical degradation, and does not mind emotional atrophy.

The threats to human needs and susceptibilities that are easiest to identify and control are those resulting directly from physicochemical and biological factors of the environment—such as unhealthful climate, pollution, disease germs, crowding, various forms of excessive stimuli. But the influences that affect human life most profoundly are not always the direct ones measured in such objective values as temperatures, chemicals, decibels, kinds of germs, or numbers of people. Man converts all the things that happen to him into symbols, then commonly responds to the symbols as if they were actual external stimuli. All perceptions and interpretations of the mind become so profoundly translated into organic processes that the actual biological and mental effect of a stimulus commonly bears little resemblance to the direct effects which could have been expected from its physicochemical nature.

Since each stimulus has a symbolic quality often more important than its objective characteristics, the effective environment does not consist only of the external forces and substances that impinge on the organism at a given time, but includes also the genetic, social, and individual memories of related past experiences. Mankind's responses to the environment always involve the biological remembrances of the past which in turn condition aspirations for the future. What man becomes is thus largely determined by the adjustments he has made to the stimuli he has experienced.

Man can learn to tolerate treeless avenues, starless skies, tasteless food, a monotonous succession of holidays which have become spiritless and meaningless because they are no longer holy days, a life without the fragrance of flowers, the song of birds, the joyous intoxication of spring, or the melancholy of autumn. Loss of the amenities of life may have no obvious detrimental effect on man's physical well-being or on his ability to perform effectively as part of the economic or technological machine. Increasingly, in fact, most professional activities are carried on in denatured dwellings, offices, and industrial plants. The popularity of the Los Angeles airport for important meetings has already been mentioned. Schools for young children are being built underground to facilitate the upkeep of rooms and lessen distractions for the pupils!

Little if anything is known, however, of the ultimate effect on man of such drastic elimination of the natural stimuli under which he has evolved as a biological being. Air, water, soil, and fire, the rhythms of nature and the variety of living things, are not of interest only as chemical mixtures, physical forces, or biological phenomena; they are the very influences that have shaped human life and thereby created deep human needs that will not change in the foreseeable future.[15] The pathetic weekend exodus to the country or beaches, the fireplaces in overheated city apartments, testify to the persistence in man of biological and emotional hungers that developed during his evolutionary past, and that he cannot outgrow.

Man will continue searching for significance by relating himself to other men, and to the totality of the universe that he may identify with God. But while pursuing significance outside of himself he should not forget that he is still of the earth. Like Antaeus of the Greek legend, he loses his strength when both his feet are off the ground.

Our Buildings Shape Us

Many persons are becoming weary and frazzled by the rat race of constant change. Adults are exhausted by the struggle and teenagers do not find it worthwhile. As they watch the maddening complexity of life, and the frantic efforts to invent new technologies for solving the problems that technology itself has created, their cry goes up, "Stop the world—I want to get off." They want to go back to Arcadia in the hope of recapturing the simplicity and purity of original life. But Arcadia has never existed except in our dreams. Furthermore, we could not re-enter Arcadia even if it did exist, encumbered as we are by the biological and social memories of civilized life.

When Captain James Cook, Louis Antoine de Bougainville, and their successors achieved their first contacts with the Polynesian world in the eighteenth century, they thought they had finally discovered primitive life uncontaminated by the artificialities of civilization. Little surprise that sailors, fed on salt beef and hard biscuit, living on the edge of scurvy, and sexually starved, should have thought themselves in paradise on reaching the green and balmy South Seas shores and accepting the amorous welcome of the Polynesian women. They failed to notice the shortcomings and tragedies of Polynesian life because their judgment had been warped by the romantic philosophy of naturalism then prevailing in the intellectual circles of Europe. The magic of their contemporary Jean Jacques Rousseau's message in particular had prepared them to believe that the noble happy savage lived in undisturbed nature and never experienced any problems.

In reality, any form of organized social life (and all human life is socialized) has its own brand of restrictions, conflicts, and frustrations, added to the problems that each individual person experiences in meeting day-to-day requirements. To live is to struggle. A successful life is not one without ordeals, failures, and tragedies, but one during which the person has made an adequate number of effective responses to the constant challenges of his physical and social environment.

Today, as in the time of Cook and Bougainville, most of us at some time yearn to escape to some Friendly Isle and recapture the quality of primitive life. But this we cannot do, partly because the human adventure has implied from the beginning an irreversible dependence on the social and technological management of nature, and more importantly because the past is incorporated in the present.

One of the most eloquent warnings that we cannot go back to Arcadia came from the English writer D. H. Lawrence (1885–1930), who also thought for a while that he could overthrow his complicated past and adopt the unspoiled way of life among the friendly people of the South Seas. "There they are, these South Sea Islanders, beautiful big men with their golden limbs and their laughing, graceful laziness. . . . They are like children, they are generous: but they are more than this. They are far off, and in their eyes is an early darkness of the soft, uncreate past. . . . There is his woman, with her knotted hair and her dark, inchoate, slightly sardonic eyes. . . . She has soft warm flesh, like warm mud. Nearer the reptile, the Saurian age. . . .

"Far be it from me to assume any 'white' superiority. It seems to me, that in living so far, through all our bitter centuries of civilization, we have still been living onwards, forwards. . . . The past, the Golden Age of the past—what a nostalgia we all feel for it. Yet we don't want it when we get it. Try the South Seas."[16]

Just as we are earthbound forever by our physiological dependence on the earth's atmosphere, so we are bound to our own times because the needs we have developed through 10,000 years of civilization can no longer find satisfaction in the "darkness of the soft, uncreate past."

While the explorers of space and of ocean depths are struggling to duplicate the terrestrial environment in their capsules, physiologists, psychologists, architects, and city planners are anxiously trying to formulate an environment optimum for man on earth. Most demographers and sociologists believe that, barring a world catastrophe, the great majority of people all over the world will soon live in urban agglomerations that will eventually extend over hundreds of miles. Here is the picture for the United States as seen by one of the American leaders in the field of planning and development: ". . . half a century from now the population of the United States will exceed 500 million people. If this proves true, we will have 500 to 1,000 metropolitan areas as now defined. The largest of these, the East Coast New York region, will contain perhaps 50 million people and will be part of a larger region exceeding 100 million people. There may be several other metropolitan areas with 25 million population. Presumably 85 to 95 percent of the population will be urban."[17]

In the past the basic population lived in the country or in small towns. Because of the enormous increase in agricultural productivity, people now move away from the country even while the total population is becoming larger. Urbanism is the way of life for an overwhelming majority of people, including not only the residents of the compact city, but also the suburbanites and the exurbanites. For all of them, urbanism implies that practically all individual and social activities are influenced by technology. Scientists must therefore concern themselves with the immediate and long-range effects of technology on human life.

The myth has grown that because man has an infinite capacity to adapt to changing environments, we can endlessly and safely transform his life and indeed himself by technology. In reality, there are biological and psychological limits to man's adaptability, and these should determine the frontiers of technological change.

Another myth states that, by proper study, it should be possible to define an environment having characteristics optimum for human life. This is impossible because men differ in their tastes and aspirations and therefore have different environmental needs. To realize the multifarious biological and spiritual potentialities of mankind requires an immense diversity of environments. The real problem therefore is to learn how environmental forces can be best managed to foster the various manifestations of happiness and creativity in mankind. Technology should have as its most important goal the creation of environments in which the widest range of human potentialities can unfold.

Knowledge of environmental effects is most precise and extensive with regard to the needs and responses of vigorous young men functioning under extreme and unusual conditions. For obvious reasons, much physiological and psychological research relevant to these problems has been sponsored by the armed forces, the space agencies, and the civil-defense program. This kind of research deals especially with operational situations in which it is essential to know the thresholds and limits of physical and psychical performance and endurance. The problems of ordinary life, however, are very different from those studied by specialized agencies.

They are more complex, and especially less well defined, because normal human populations are extremely heterogeneous and rarely live under extreme environmental situations. In ordinary life, furthermore, the criteria of health, well-being, comfort, productivity and happiness do not lend themselves readily to scientific statements because they are highly subjective and socially conditioned.

A few very broad generalizations can serve as a theoretical basis for discussing the effects of the environment on human life. Ideally, all aspects of the environment should form an integrated ecological system in which the welfare of any part of the system is dependent upon the welfare of all the others. In the light of ecological theory, man is part of the total environment and therefore cannot achieve and maintain physical and mental health if conditions are not suitable for environmental health. For this reason, it is ecologically and indeed logically impossible to define an optimum environment if one has only man in mind.

Another generalization, popularized by Toynbee, is that the type of environment most conducive to human development is one sufficiently changeable to pose constant challenges but not so severe as to prevent successful responses.[18] The temperate regions seem to be most satisfactory from this point of view; in fact, the Yale University geographer Ellsworth Huntington thought that the Connecticut area in which Yale University is located provides an ideal environment for the flowering of civilization.[19] Wherever life is without challenge and too comfortable, as supposedly in the Polynesian islands, the best that can be hoped for is an arrested or static civilization rather than one that is innovating and on-going.

The challenge-and-response theory teaches that the

environment should provide the proper intensity and variety of stimulation. In addition, there are a few environmental imperatives that derive from the unchangeable aspects of man's nature. Best understood among these imperatives are the determinants of health and disease so extensively studied by the biomedical sciences. Even this aspect of the environmental problem is less well defined than used to be thought, however, because the patterns of disease change rapidly and unpredictably with the conditions of life. Positive and absolute health has proved so far to be a constantly receding mirage.[20]

The interrelationships between human beings are naturally among the most important of the factors to be considered by planners, but little is known concerning real human needs in this regard. Many anthropologists and sociologists have taken a gloomy view of the effects that the modern conditions of life have on human relationships; they see in the present scene little chance to satisfy the essential need for the intimate kind of contact with a very few persons that can occur only within a small primary group. "In the old society, man was linked to man; in the new agglomeration—it cannot be called a society—he is alone. . . . All the evidence of psychiatry shows that membership in a group sustains a man, enables him to maintain his equilibrium under the ordinary shocks of life, and helps him to bring up children who will in turn be happy and resilient. . . . The cycle is vicious; loss of group membership in one generation may make men less capable of group membership in the next. The civilization that, by its very process of growth, shatters small group life will leave men and women lonely and unhappy."[21]

Some urban planners advocate social and architectural arrangements which provide each individual person with three or four intimate contacts at every stage of his existence.

But the great mobility of our populations, the high levels of crowding, and the increased complexity of social life make such intimate associations almost impossible for the larger percentage of the public. From all points of view, population pressure probably constitutes the most important single handicap to creating urban environments with proper biological qualities. There is no immediate danger that the United States will experience shortages of food or even a decrease in the economic standard of living as a result of population pressure. Nevertheless we suffer from overpopulation because human life is affected by determinants that transcend technology and economics.

The greatest dangers of overpopulation come paradoxically from the fact that human beings can make adjustments to almost anything. Congested environments, even though polluted, ugly, and heartless, are compatible with economic growth and with political power. Similarly, social indifference, aggressive behavior, or the rat race of overcompetitive societies will not necessarily destroy mankind. Crowding, however, can damage the physical and spiritual qualities of human life through many mechanisms, such as the narrowing of horizons as classes and ethnic groups become more segregated, with the attendant heightening of racial conflicts; the restrictions on personal freedom caused by the constantly increasing need for central controls; the deterioration in professional and social services; the destruction of beaches, parks, and other recreational facilities; the spreading of urban and suburban blight; the traffic jams, water shortages, and all forms of environmental pollution.

We do not recognize danger in crowding as long as we can produce enough food for physical growth and enough goods for economic growth. We do not sense the evil because we regard ourselves, and other men, as things rather than as fellow human beings. The availability of food, natural resources, and power required for the operation of the body machine and of the industrial establishment is not the only factor to be considered in determining optimum population size. Just as important for maintaining the quality of *human* life is an environment in which it is possible to satisfy the longings for quiet, privacy, independence, initiative, and open space. These are not frills or luxuries; they constitute real biological necessities. They will be in short supply long before there are critical shortages of energy and materials to keep the human machine going and industry expanding.

In theory, all human beings have the same essential needs, but in practice actual needs are socially conditioned and therefore differ profoundly from one human group to another. Even food requirements cannot be defined without regard to the social context. The value of an article of food is not determined only by its content in protein, carbohydrate, fat, vitamins, minerals, and other chemical components. A particular food has in addition symbolic values which make it either essential or unacceptable, depending upon the past experiences of the consumer.[22] These symbolic aspects of nutrition are of importance not only among primitive people. Americans are even more reluctant to eat horse meat than Frenchmen are to eat cornbread.

The kinds of technical equipment needed also vary with time and from place to place. The ancient Mayas created an extraordinarily sophisticated culture and marvelous monuments of imposing size without using the wheel as a means of transportation, although, as some of their toys show, they knew the wheel and its possible applications. Similarly, the wheel was not used by the tribes of Central Africa until modern times, even though representations of it in Neolithic paintings have been discovered. In Central America and Africa other forms of transportation were more practical until the white man had cut wide roads through the equatorial forest. Human need is not a fixed quality. As stated by Gordon Childe:

"No doubt the efficiency of an automobile to satisfy the need for transport under specific conditions can be determined with mathematical accuracy. But is man's need for transport a fixed quantity in any real sense? Did a reindeer hunter in 30,000 B.C., or an Ancient Egyptian in 3000, or an Ancient Briton in 30, really need or want to travel a couple of hundred miles at 60 m.p.h.?

"To a Magdalenian society in the last Ice Age a harpoon of antler was just as efficient as a steam trawler is today. With the former, tiny groups could get all the fish they needed. . . ."[23]

Similarly, needs that appear vital today may become trivial in another generation, not because man's biological nature will change, but because the social environment will not be the same. For example, the individual motor car may progressively disappear if, as is probable, driving loses its appeal and if, as we may hope, people have more uses for their leisure time within walking distance of their houses. The individual, detached house may also become obsolete once home ownership loses its symbolic meaning of economic and social independence by reason of more generalized prosperity and financial security. A new generation may learn again to prefer the excitement of the compact city to the bourgeois comfort of the suburb and thus bring to an end the lawn-mower era.

Henry David Thoreau spent a year in a hut that he

built for himself near Walden Pond to demonstrate that the essential needs of man are small. But he had taken along a number of scholarly books, and plenty of paper to write his diary. His biological needs were small indeed, but he had wants that he shared with his Concord circle.

The phrase "essential need" is therefore meaningless, because in practice people need what they want. Needs are determined less by the biological requirements of *Homo sapiens* than by the social environment in which a person lives and especially that in which he has been brought up. The members of a given social group generally come to desire, and consequently develop a need for whatever is necessary for acceptance in the group. The good life is identified with the satisfaction of these needs, whatever their biological relevance.

Wants become needs not only for individual persons, but also for whole societies. Monuments dedicated to the Virgin Mary and the saints were apparently a need for thirteenth-century Europe, which devoted an enormous percentage of its human and economic resources to the creation of churches and monasteries that appear to us extravagant in relation to the other aspects of medieval life. In our times, the Great Society seems particularly concerned with creating a middle-class, materialistic civilization with a veneer of uplifting platitudes. This concern also creates special needs, including frozen fruit juice for breakfast, a different dress for every day at the office, a playroom in the cellar, and a huge lampshade in front of the picture window.

The environment men create through their wants becomes a mirror that reflects their civilization; more importantly it also constitutes a book in which is written the formula of life that they communicate to others and transmit to succeeding generations. The characteristics of the environment are therefore of importance not only because they affect the comfort and quality of present-day life, but even more because they condition the development of young people and thereby of society.

The view that man can shape the future through decisions concerning his environment was picturesquely expressed by Winston Churchill in 1943 while discussing the architecture best suited for the Chambers of the House of Commons. The old building, which was uncomfortable and impractical, had been bombed out of existence during the Second World War. This provided an opportunity for replacing it by a more efficient one, having greater comfort and equipped with better means of communication. Yet Mr. Churchill urged that the Chambers should be rebuilt exactly as they were before. In a spirited speech, he argued that the style of parliamentary debates in England had been conditioned by the physical characteristics of the old House, and that changing its architecture would inevitably affect the manner of debates and, as a result, the structure of English democracy. Mr. Churchill summarized the concept of interplay between man and the total environment in a dramatic sentence that has validity for all aspects of the relation between human life and the environment: "We shape our buildings, and afterwards our buildings shape us."[24]

While the total environment certainly affects the way men feel and behave, more importantly it conditions the kind of persons their descendants will become, because all environmental factors have their most profound and lasting effects when they impinge on the young organism during the early stages of its development.

Mr. Churchill was promoting a conservative policy—

the maintenance of traditional parliamentary practices—when he developed his argument that our buildings shape us. Most educational and social systems also try to force the young into traditional patterns through environmental manipulations, and despite appearances they largely succeed. Americans, Englishmen, Frenchmen, Germans, Italians, or Spaniards acquire their national characteristics because they are shaped during early life by their buildings, educational systems and ways of life. But such shaping need not be only for the preservation of the past. It can be oriented toward the future.

The Israeli *kibbutz* has demonstrated that a systematic program of child-rearing in collectives can, in a single generation, give to children a healthy and vigorous personality entirely different from that of their parents.[25] The success of this social experiment does not establish the desirability of the results that it achieved, but it does show that the environment can be used to alter persons and institutions as well as to preserve the *status quo*.

Any manager of television or of any other publicity program directed to the young must often be tormented, if he has any social conscience, at the thought that he is conditioning the tastes, opinions, and reaction patterns of his audience lastingly and perhaps irreversibly. Would that there were a Winston Churchill of the publicity profession capable of conveying to his colleagues the biological law that: We shape our programs, and then afterwards our programs shape us and our children!

Educators have long known that all aspects of the environment affect the unfolding of human potentialities and the character of their manifestations. Political leaders have also used this knowledge to manipulate public opinion and especially to shape the minds of the young. A broader and more hopeful aspect of the same problem is the possibility of creating conditions that will enable individual human beings to discover what they are capable of becoming and to enjoy the freedom of making their lives what they want them to be.

There certainly exist in the human genetic pool rich potentialities that have not yet been fully expressed and that would permit mankind to continue evolving socially if conditions were favorable for their development. The diversity of civilizations originates from the multifarious responses that human groups have made in the past and continue making to environmental stimuli. This versatility of response, in turn, is a consequence of the wide range of potentialities exhibited by human beings. Since most of these potentialities remain untapped, each one of us actually becomes only one of the many persons he could have been.

Human potentialities, whether physical or mental, are expressed only to the extent that circumstances are favorable to their manifestation. The total environment thus plays a large role in the unfolding of man's nature and in the development of the individual personality.

In practice, the latent potentialities of human beings have a better chance to come to light when environment provides a variety of stimulating experiences, especially for the young. As more persons find the opportunity to express a larger percentage of their biological endowment under diversified conditions, society becomes richer and civilizations continue to unfold. If surroundings and ways of life are highly stereotyped, the only components of man's nature that flourish are those adapted to the narrow range of prevailing conditions.

In theory, the urban environment provides a wide range of options. The present trends of urban life are usually assumed to represent what people want, but in reality the trends are determined by the available choice. While people need what they want, what they want is largely determined by the choices readily available to them. It has been said that children growing up in some of the most prosperous American suburbs may suffer from being deprived of experiences. In contrast, the Lower East Side of New York City in the 1900s, despite its squalor and confusion, provided one of the richest human environments that ever existed. Children there were constantly exposed in the street to an immense variety of stimuli from immigrants of many cultures; many of these children became leaders in all fields of American life.[26]

In any case there is no doubt about the sterilizing influences of many modern housing developments, which, although sanitary and efficient, are inimical to the full expression of human potentialities. Many of these developments are planned as if their only function was to provide disposable cubicles for dispensable people.

In *The Myth of the Machine*, Lewis Mumford states that "if man had originally inhabited a world as blankly uniform as a 'high-rise' housing development, as featureless as a parking lot, as destitute of life as an automated factory, it is doubtful that he would have had a sufficiently varied experience to retain images, mold language, or acquire ideas."[27] In this statement, Mumford had in mind the emergence of man's attributes during evolutionary times. He would probably be willing to apply the same concepts to modern life. Irrespective of their genetic constitution, young people raised in a featureless environment and limited to a narrow range of living experiences are likely to suffer from a kind of deprivation that will cripple them intellectually and emotionally.

Man has been highly successful as a biological species because he is adaptable. He can hunt or farm, be a meat-eater or a vegetarian, live in the mountains or by the sea-shore, be a loner or a team-member, function in a democratic or totalitarian state. History shows, on the other hand, that societies which were efficient because they were highly specialized rapidly collapsed when conditions changed. A highly specialized society, like a narrow specialist, is rarely adaptable.

Cultural homogenization and social regimentation resulting from the creeping monotony of overorganized and overtechnicized life, of standardized patterns of education, mass communication, and entertainment, will make it progressively more difficult to exploit fully the biological richness of our species and may handicap the further development of civilization. We must shun uniformity of surroundings as much as absolute conformity in behavior and tastes. We must strive instead to create as many diversified environments as possible. Richness and diversity of physical and social environments constitute an essential criteria of functionalism, whether in the planning of cities, the design of dwellings, or the management of individual life.

Diversity may result in some loss of efficiency. It will certainly increase the variety of challenges, but the more important goal is to provide the many kinds of soil that will permit the germination of the seeds now dormant in man's nature. Man innovates and thus fully expresses his humanness by responding creatively, even though often painfully, to stimuli and challenges. Societies and social groups that have removed themselves into pleasure gardens where all was designed for safety and comfort have achieved little else and have died in their snug world.

Many animal species other than man create buildings and institutions that complement the biological attributes of their bodies and serve as a focus of organization for their life. The beehive, the decorated nest of the Australian bower bird, the dam and house of the beaver are but a few examples illustrating that animals can organize inert materials to create new environments which are the equivalent of the institutions men create out of nature. The artificial environments created by animals often have esthetic quality because they are built with great economy of means, are designed to fit their purpose exactly, and enhance the relation of the animal to the rest of nature. Many of the institutions created by preindustrial people have much the same characteristics and qualities. The medieval village, the Italian hill town, the seventeenth-century New England village illustrate that human institutions can also make use of inert materials and topographical characteristics to establish harmonious relationships between man and nature.[28]

All great forms of human architecture incorporate an attitude toward life. This is true of the home, the garden, the temple, the village, and the metropolis. To be successful esthetically and practically, buildings and other artifacts must reflect the spirit of the institutions from which they originate. Whenever societies have formulated worthwhile thoughts or attitudes, artists have been forthcoming to give them a vivid and appropriate physical form. The Parthenon, Chartres Cathedral, the Renaissance cities symbolize whole civilizations. In our times the great bridges symbolize man's desire to span the gaps that separate him from other men. Brooklyn

Bridge was immediately acclaimed throughout the world, and still constitutes, to my taste, one of the marvels of the modern era. It demonstrated that steel and concrete could serve for the creation of meaningful beauty; through it, technology in the service of a purpose became a joy and inspiration for painters and poets.

Architectural form at its best has always been an expression of the ideals and underlying social philosophy of human institutions; it constitutes an organic structural expression of social need. The German art critic Wolfgang Braunfels recently illustrated this thesis with a number of telling examples: The simple farmhouse represents the family maintaining itself by working on the land. The Carolingian Benedictine monastery incorporates the rule of Saint Benedict, according to which the monks not only worked, ate, and prayed together but also walked in slow-moving processions to the different functions of the day. The city-state of Siena in the thirteenth and early fourteenth century, rich in economic wealth and art, governed by a democratic assembly and citizens' committees, tried at every stage of its development to make the city a mirror of the entire cosmos and of life itself. The Palazzo Farnese, built by Pope Paul III for his sons and nephews in the heart of sixteenth-century Rome, symbolized the greatness of the family and served as a proper setting for the famous Farnese collection of art. The Palais de Versailles was built and designed not only for the personal aggrandizement of the Roi-Soleil, Louis XIV, but also to symbolize his unique position in the state.[29]

Two of these building types, the farmhouse and the monastery, evolved slowly from crude and humble beginnings in the course of several generations. In contrast, the Palazzo Farnese and the Palais de Versailles emerged fully developed, functionally and artistically, from the minds of men with powerful visions. Both these methods of growth operated in Siena, which illustrates at every period of its history a complex interplay between natural environment, social institutions, and the views men form of themselves and of the world. Great architecture and great planning never develop in an intellectual or spiritual vacuum. Plans and buildings express the spirit of the social institutions from which they arise, and then they influence the further development of society.

Several modern factories achieve genuine functional beauty because they express in their design the objectives of great precision, maximum output, and minimum cost, which are the ideals of modern technology. Many great bridges and a few highways are among the most notable achievements of our times, not so much because of the technological skills their construction required, but because they have the larger significance of expressing the compelling desire of modern man to explore and expand his personal world.

Unfortunately, most apartment and office buildings have nothing to communicate beyond efficiency and conspicuous wealth, hence their architectural triviality. As to our cities, no planning will save them from meaningless disorder leading to biological decay, unless man learns once more to use cities not only for the sake of business, but also for creating and experiencing in them the spirit of civilization.[30] We have inherited countless great monuments from the past. The automobile seems to be our most likely bequest to future generations; they will have to retrieve it from junkyards. The automobile is the symbol of our times and represents our flight from the responsibility of developing creative associations with our environment for the sake of the future.

Our dismal failure to develop really desirable cities,

offices, and dwellings is not due to deficient engineering or bad workmanship but to the fact that technological skill cannot create anything worthwhile if it does not serve a worthwhile purpose. Our institutions are not really designed to help in developing the good life, but rather to make human beings more productive and more efficient tools of industry and commerce. Yet it is obvious that productivity and efficiency have no value in themselves; they have merit only as means to ends. In fact, excessive concern with productivity and efficiency interferes with the pursuit of significance.

Civilizations are like living organisms; they evolve according to an inner logic that integrates their historical determinants, their natural resources, and their acquired skills. It is also true, however, that most civilizations have suffered and many have died when this logic generated undesirable trends from which they could not or did not try to escape.

The logic of medieval thought led into scholastic verbiage; Gothic architecture toppled when it tried to outdo itself in the high towers of Beauvais; the scholarly learning of the nineteenth century is now degenerating into dehumanizing specialization. Scientific technology is presently taking modern civilization on a course that will be suicidal if it is not reversed in time. What, for example, will be the ultimate consequence for the United States of the fact that three centuries of homesteading, coupled with a national tradition of compulsive nomadism, has imposed on the overwhelming majority of urban dwellers the desire to occupy a one-family house, to drive to work in a private automobile, and to identify leisure time with essentially aimless movement. Such compulsions call to mind certain biological trends that have brought about the extinction of countless animal species.

It is a truism that technological advances do not determine what is desirable but only what is feasible at a given time. We shall not improve the quality of life and of the environment merely by developing greater technological skills. In fact, as stated by Norbert Wiener, nothing will make the automated factory work automatically for human good, unless we have determined worthwhile ends in advance and have constructed the factory to achieve these ends.[31] In principle, nobody wants nastiness or ugliness, and everybody is for improvement of life and of our surroundings. In fact, however, our communities want to possess things and to engage in activities that are incompatible with civilized ways of life and pleasant surroundings.

Many times in the past, civilizations have lost the will or the ability to change after they have set on a certain course. Such civilizations soon exhaust the spiritual content and creativeness that characterized their initial phase. They usually retain for a while a certain kind of vigor based on orthodox classicism but soon degenerate into triviality before foundering in the sea of irrelevance.

If it is true, as it appears to be, that our environment and way of life profoundly affect our attitudes and those of following generations, nothing could be more distressing for our immediate and distant future than the decadence and ugliness of our great urban areas, the breakdown in public means of transportation, the overwhelming accent on materialistic and selfish comfort, the absence of personal and social discipline, the sacrifice of quality to quantity in production as well as in education. The lack of creative response to these threats is particularly discouraging because all thinking persons are aware of the situation and are anxious to do something to correct it. But common action cannot be mustered because it demands a common faith that does not exist.

It is because we need a common faith that the search for significance is the most important task of our times.

Outgrowing the Growth Myth

The pursuit of significance seems a futile chase when judged in the perspective of history. Every time mankind has approached an ideal that gave significance to life, this ideal has vanished like a will-o'-the-wisp. Many kinds of religious, philosophical, and social faiths have appeared in the past and illuminated the human condition for a while, only to be lost in a morass of philosophical uncertainties and hair-splitting arguments.

During the Middle Ages, Christianity acted as a great unifying force by giving the people of Europe a few common aspirations and social disciplines derived from the love and fear of God. Christian mysticism mobilized human energies for spectacular collective tasks such as the building of the Romanesque and Gothic monasteries and cathedrals. Mont-Saint-Michel and Chartres, as Henry Adams perceived, cannot be fully understood or enjoyed unless we sense that each stone was quarried, moved, and chiseled by the collective efforts of noblemen and peasants, architects and sculptors, all united in the worship of Saint Michael and the Virgin Mary.[32] Progressively, however, Christians became involved in repetitious theological arguments; from a mystic doctrine of love, Christianity evolved into conservative and uninspired orthodoxy. Now it often degenerates still further into vague social ethics; theologians engage in spurious philosophical discussions to reconcile Christianity with the meaningless statement that God is dead.

Beginning with the Enlightenment in the eighteenth century, scientific rationalism increasingly gained ground as the unifying faith of mankind. During the past few decades, however, it too has begun to lose its force because its intellectual and practical limitations are becoming evident. Hardly anyone doubts that science is the most powerful force in the modern world, but there are few persons, especially among sophisticated scientists, who still believe that it can explain the riddle of the universe or alone give direction and significance to human life.

The boundaries of modern science will certainly be determined by limitations to man's understanding inherent in man's nature. At the edge of biology, for example, we encounter the chasm between the phenomena of life that objective science can describe and the subjective experiences known only by the mind. The physical sciences also present us with insoluble contradictions when we try to comprehend the limits of space or the beginnings of time. Furthermore, scientific achievements commonly raise ethical issues that many scientists consider outside their professional competence. Science and technology, they point out, are tools and instruments which are in themselves amoral and can be used either for the benefit or detriment of mankind. Even the faith that science can solve most practical problems has been recently tarnished by the increasing awareness that scientific technology endlessly creates new problems as it solves old ones.

Since time erodes all religious, philosophical, and social faiths, is it at all reasonable to search for meaning in life, to hope for perfect solutions to practical issues? The difficulty in dealing with the problem of significance may come from the fact that we ask the wrong questions and misinterpret the successes which mankind has had in achieving periods of unifying faith.

The philosophers of the classical world, the builders of cathedrals, the founders of scientific rationalism, or the leaders of any of the social and religious movements that have shaped history were not primarily concerned with formulating well-defined philosophies or solving immediate practical problems. What the various forms of faith that gave a certain unity to groups of people had in common was rather a set of values that made it possible for man to transcend his own individual life and to find significance in a much larger context. Whether based on religious, philosophical, or social convictions, the feeling of significance derives from man's awareness, vague as it may be, that his whole being is related to the cosmos, to the past, to the future, and to the rest of mankind. Such a sense of universal relatedness is probably akin to religious experience.

Faith has taken many forms in the course of history, but whether symbolized in a personal God, or conceived as an abstract philosophy, it always involves a view of man that extends beyond the here and now. The more extensive the relatedness the greater the significance; this is why religion and philosophy, despite their apparent lack of practical value, retain such appeal for mankind.

The purely biological study of man deals with such problems as: How and where did *Homo sapiens* evolve? What are the structures and mechanisms through which his body and mind operate? What are the determinants of his behavioral and social patterns? To what extent is he influenced by his surroundings and ways of life?

The answers to all these questions are essential for the understanding of *Homo sapiens* and for the better management of human life, but they throw little light on some of the problems that have always preoccupied healthy human beings. While it is essential to know as much as possible concerning man's origin, his development, his biological and social mechanisms, it might be even more important to help each individual person understand where he belongs in the cosmic order, and what gives significance to his own life. Religions and philosophies have contributed little if anything to the understanding and improvement of man as a living machine; they have, nevertheless, helped him immeasurably by providing hypotheses and tentative answers to the haunting questions: Where do I come from? Where am I going? And especially, who am I? The tragedy of the juvenile delinquent gives a special poignancy to these questions.

Typically, the true juvenile delinquent does not behave antisocially out of deliberate wickedness. He acts for the immediate satisfaction of an urge, an appetite, or a whim. He lives only in the present. For congenital and more often sociocultural reasons, he is incapable of relating himself to others, to the past, or to the future. The worst aspect of his fate is that he finds no significance in life and therefore has no reason to develop a sense of responsibility.

It would be unreasonable and unfair to assert that ordinary human beings now behave as juvenile delinquents. Nevertheless, most of them in the countries of Western civilization, particularly in present-day America, must be considered delinquent because they act as if the immediate satisfaction of all their whims and urges were the only criteria of behavior, without regarding the consequences for the rest of nature and for posterity. Textbooks damn Louis XV for his irresponsible remark, "*Après moi, le déluge.*" Yet we too are using the earth as if we were the last generation. Socially, we behave as if we were willing to excuse our misdeeds with the question "What has posterity done for me?"

All successful individual lives, and all successful civilizations, have been supported by an orderly system of relationships linking man to nature and to society. These relationships, which are absolutely essential not only to the well-being of the individual person but also to the survival of human groups, are now rapidly and profoundly disturbed by modern life. At stake, therefore, is not only the rape of nature but the very future of mankind.

Before illustrating the destructive forces set in motion in the modern world, let me emphasize that the guilt for the present frightening social and environmental situation cannot be placed on villains with selfish interests, bent on doing harm to mankind. In fact, the guilt cannot be placed on any particular person. For example, the knowledge of ionizing radiations and of nuclear energy was developed by men with such exalted ideals that they might be regarded as modern saints, yet immense harm has come from their scientific activities. The internal-combustion engine was at first a boon to mankind; yet the overuse of the automobile is polluting the air, disorganizing life, and destroying cities. In other words, the dangers created by technological and social innovations do not come from conscious human ill will but from the fact that our political and social mechanisms are outdated and out of keeping with the modern world; they can neither predict nor control the nefarious consequences of the exploitation of technological developments for economic purposes.[33]

As long as social and environmental changes were

slow and few in number, mankind could take them in stride through biological and social adjustments which prevented irreparable damage to nature and man. But now too much is happening too fast. Biologically and socially, the experience of the father is almost irrelevant to the conditions under which his son will live and can no longer serve as a dependable guide for judgment and action.

The world population has been increasing more or less steadily ever since the appearance of man on earth, but the rate of increase has accelerated so much during the past century that a critical point is about to be reached. Man has slowly become adjusted to the changes necessitated by larger populations, but we are approaching the threshold of safety with regard to both the speed and the extent of social and technological innovations. There is no danger for the Eskimo in letting the smoke out of his snow house; he does not need to worry about where the smoke goes after it escapes. But smoke disposal becomes a critical problem when millions of people are concentrated in a small area. We can expect worldwide disasters unless the population is rapidly stabilized and unless further transformations of environments and ways of life are properly managed. We can no longer dismiss environmental problems as the Eskimo could in isolated Arctic settlements.

Countless innovations are being introduced almost simultaneously; they reach into all parts of the world and affect practically all aspects of life before anything is known of their potential effects on the human organism, on social structures, or on environmental conditions. It was 100,000 years before the rough Chellean hand ax was replaced by the smoother Acheulian tool during Paleolithic times; the horse remained the fastest means of transportation until the middle of the nineteenth century; the speed of railroads has hardly increased in the past hundred years. In contrast, we have moved in one generation from railroad and automobile speed to that of the propeller plane, then of the supersonic plane. The techniques of food production and distribution had remained much the same since Neolithic times; now suddenly farmers and food processors are using thousands of chemicals which find their way into vegetable and animal products and thus into the human body.

The Industrial Revolution, with mass production of energy and manufactured goods and the rapid injection of technological procedures into all social activities, is beginning everywhere to disrupt the great dynamic processes which have so far maintained the earth in a state compatible with human life. Even agriculture is now industrialized; the massive use of chemical fertilizers and pesticides is changing the nature of the land, spoiling water resources, and creating violent ecological disturbances.

Increasingly, we cut down forests and flood deserts to create more farmland. On the other hand, we destroy fertile agricultural fields to build factories, highways, and housing developments, without regard to natural and historical scenery. We first cleared the forests to make way for the farms, then we cleared the farms to accommodate the cities and their suburbs. Almost everywhere, the land is being used not as a home, not as an environment for the creation of human culture, but as a source of exploitation and speculation.

We eliminate all forms of wildlife that compete with us for space and for food; we tolerate animals, plants, and landscapes only to the extent that they serve economic purposes. It is perhaps symbolic that rats appear to be the only mammals that have increased in numbers during the past century as much as men.

We are running out of breathable air in many cities, and on occasion leafy vegetables have been so affected by pollutants that blemishes have rendered them unmarketable—a symbol of what air pollution does to all living things.

Most apparent has been the destruction of great water systems. Lake Erie has been turned into a cesspool. The Ohio, the Hudson, the Merrimac, and countless other streams are used as sewers. Billions of dollars will have to be spent for correcting the damage done by water pollution, and in certain cases the point of no return may have been reached.

Because of the population increase, and even more because of thoughtless and wasteful demands, all natural resources will soon have to be used for strictly utilitarian ends. Disruption of the water cycle is speeding water on its way to the sea and increasing its destructive action on land surfaces; denudation of the soil is creating dust bowls; pollution of air and water is beginning to upset the biological balance and to damage human health. Man is rapidly destroying all the aspects of the environment under which he evolved as a species and which created his biological and emotional being.

The medical sciences have developed methods so powerful that they can affect simultaneously and in an unpredictable manner the fate of immense numbers of people and their descendants, often creating new pathological processes as they control old diseases. There is no historical precedent for such a massive medical intervention, nor is there enough theoretical knowledge of population genetics and of physiology to predict the nature or magnitude of the effects that these revolutionary changes will have on the constitution and general resistance of man in the future.[34]

There is nothing fundamentally new in the fact that civilization alters man, nature, and their interrelationships. For many thousands of years, man has modified his environment and consequently himself by using fire, farming the land, building houses, opening roads, killing or domesticating animals, and controlling his own reproduction.

Primitive people generally succeeded in stabilizing their population level.[35] The Ebers papyrus (1550 B.C.), which is the oldest-known Egyptian compendium of medical writings, contains the formula for a tampon medicated to prevent conception and this has been proved to have definite effectiveness. When Francis Place launched the birth-control movement in 1822, he listed several of the contraceptive techniques and other methods of conception control still in use today. Furthermore, abortion has always been widely practiced.

In *Critias*, Plato describes how Attica had become a "skeleton of a body wasted by disease," because the forests had been cut down. Overgrazing had added to the damage of deforestation, drying up the springs, and destroying the most fertile soils because the water was lost "by running off a barren ground to the sea." Pliny (A.D. 23–79) told of man's altering climates by changing river courses and draining lakes, with the result that olives and grapes were killed by frost.[36] The vegetation of Italy was transformed during Greco-Roman times by the westward dispersal of grapevines, olive trees, fig trees, stone fruits, wheat, rice, some shade trees, and many ornamental plants, and later by the addition of sugar cane, date palms, and some citrus fruits.

But all these changes occurred gradually. They cannot be compared with the massive human interventions such as the instant deforestation or the intensive monoculture which created the dust bowls and made man turn the earth, to use William Faulkner's words, "into a howling waste, from which he would be the first to vanish."

Similar remarks apply to the effects of technological changes on human life. Scientific technology today appears at first glance to be merely an extension, even though a spectacular one, of what it started out to be in the early nineteenth century. In fact, it is different in nature. Until a few decades ago scientists and technologists were concerned with well-defined problems of obvious relevance to human welfare. They saw misery and disease caused by acute shortages of food and of elementary conveniences; they observed that ignorance generated terror, superstition, and often acts of cruelty. It was urgent to abolish the threat of scarcity and to help man face the natural world without fear. Scientific technology made it possible to reach these goals and thus acted as a true servant of mankind. Unfortunately, modern man developed new technological forces before he knew how to use them wisely. All too often, science is now being used for technological applications that have nothing to do with human needs and aim only at creating new artificial wants. Even the most enthusiastic technocrat will acknowledge that many of the new wants artificially created are inimical to health and distort the aspirations of mankind. There is evidence furthermore that whole areas of technology are beginning to escape from human control. The danger from this source was forcefully stated by an eminent engineer before the American Association for the Advancement of Science in December 1966:

"Apparent overconfidence in a technical system recently permitted a catastrophic breakdown to occur in the power supply to a major concentration of the nation's industry and population. Misjudgment of the possible effects of an atomic explosion has disturbed the structure of the Van Allen belts, apparently for many years to come. The proliferation of motor vehicles, jet aircraft, and other exhaust-producing machines has contributed to massive pollution of the atmosphere. Uncontrolled disposal of industrial waste and the widespread use of chemical pesticides and weed killers have tainted water supplies and affected marine life.

"In short, the introduction of new technology without regard to *all* the possible effects can amount to setting a time bomb that will explode in the face of society anywhere from a month to a generation in the future."[37]

Technology, allowed to develop without proper control, thus may act as a disruptive force which will upset the precarious relationships upon which civilizations have been built in the past. As the English writer E. M. Forster predicted in "The Machine Stops," technology "moves on, but not on our lines; it proceeds, but not to our goals." Most of the problems posed by the use of technology are primarily social, political, and economic rather than scientific in nature. Furthermore, technology cannot theoretically escape from human control, but in practice it is proceeding on an essentially independent course, for the simple reason that our societies have not formulated directives for its control and proper use.

All societies influenced by Western civilization are at present committed to the gospel of growth—the whirling-dervish doctrine which teaches: produce more so that you can consume more so that you can produce still more. One need not be a sociologist to know that such a philosophy is insane.

Accelerated growth cannot go on for very long, let alone forever. In fact, it may be stopped earlier than anticipated by the growing awareness in the sophisticated public that uncontrolled technological growth damages the qualities of life.

In a speech entitled "Can America Outgrow Its Growth Myth," Secretary of the Interior Stewart L. Udall dared to state that man-made America can easily be considered "a catastrophe of continental proportions." He reminded his audience that "we have the most automobiles of any country in the world—and the worst junkyards. We're the most mobile people on earth—and we endure the most congestion. We produce the most energy, and we have the foulest air." He quoted with approval the mayor of Cleveland who "quipped that if we weren't careful we'd be remembered as the generation that put a man on the moon while standing knee-deep in garbage." In his peroration, Secretary Udall expressed the hope that an upsurge of anger against the destructive gospel of "growth-as-progress" would lead to more reasonable thinking and would suggest how to bring beauty and order back to our land.[38]

Eloquent and learned though they may be, speeches by government officials and university professors are never really convincing. But Secretary Udall's statements acquire additional significance from a few other signs of public resentment against the takeover of life by technology. A forerunner was the receptiveness of movie audiences to Charlie Chaplin's rebellion against the mechanized factory in the film *Modern Times.*

A rich anthology could be composed of articles and books damning or ridiculing the machine that have appeared during the past few years.

Admittedly the attitude of protest involves only a

small percentage of the general public. Most adults among us belong to the first two generations of the technological era; we naturally enjoy the luxury it has provided because it is still so new. But such enjoyment may not last long. Technological novelty soon becomes monotonous; even the space adventure is beginning to pall. As the British science analyst Ritchie Calder wrote, "The new generation was born in a world of change; they have radiostrontium in their bones; their birth rates were registered by computers and their zodiacal sign was Sputnik."[39] They are not as much impressed by technological changes as their elders are, and their own children may be even less so. Hopefully, their indifference might help America to outgrow its Growth Myth.

6.

THE SCIENCE OF HUMANITY

The Wooing of the Earth

I live in mid-Manhattan and, like most of my contemporaries, experience a love-hate relationship with technological civilization. The whole world is accessible to me, but the unobstructed view from my 26th-floor windows reveals only a confusion of concrete and steel bathed in a dirty light; smog is a euphemism for the mud that constantly befouls the sky and blots out its blueness. Night and day, the roar of the city provides an unstructured background for the shrieking world news endlessly transmitted by the radio.

Everything I eat, drink, and use comes from far away, or at least from an unknown somewhere. It has been treated chemically, controlled electronically, and handled by countless anonymous devices before reaching me. New York could not survive a week if accident or sabotage should interrupt the water supply during the summer or the electric current during midwinter. My life depends on a technology that I do not really understand, and on social forces that are beyond my control. While I am aware of the dangers this dependence implies, I accept them as a matter of expediency. I spend my days in the midst of noise, dirt, ugliness, and absurdity, in order to have easier access to well-equipped laboratories, libraries, museums, and to a few sophisticated colleagues whose material existence is as absurd as mine.

Our ancestors' lives were sustained by physical work and direct associations with human beings. We receive our livelihood in the form of anonymously computerized paper documents that we exchange for food, clothing, or gadgets. We have learned to enjoy stress instead of peace, excitement in lieu of rest, and to extract from the confusion of day-to-day life a small core of exhilarating experiences. I doubt that mankind can tolerate our absurd way of life much longer without losing what is best in humanness. Western man will either choose a new society or a new society will abolish him; this means in practice that we shall have to change our technological environment or it will change us.

The following remarks made during a discussion held at Massachusetts Institute of Technology bring out the problems posed by the adaptation of human values to technological development.

Harvey Cox: ". . . there are components of the situation which allow themselves to be addressed by technological answers. But I think there is this other one which I don't think the technological answers get to, and it has a little bit to do with a question about our basic philosophical assumptions about man, and what it means to be fulfilled."

Question from the floor: "But our basic philosophical assumptions may be pretechnological in nature, and *one of the main problems of man today may be to readjust philosophical perspective to modern technology*"[1] [italics mine—R.D.].

Adjusting man's philosophical perspective to modern technology seems to me at best a dangerous enterprise. In any case, the technological conditions under which we now live have evolved in a haphazard way and few persons if any really like them. So far, we have followed technologists wherever their techniques have taken them, on murderous highways or toward the moon, under the threat of nuclear bombs or of supersonic booms. But this does not mean that we shall continue forever on this mindless and suicidal course. At heart, we often wish we had the courage to drop out and recapture our real selves. The impulse to withdraw from a way of life we know to be inhuman is probably so widespread that it will become a dominant social force in the future.

To long for a human situation not subservient to the technological order is not a regressive or escapist attitude but rather one that requires a progressive outlook and heroic efforts. Since we now rarely experience anything directly and spontaneously, to achieve such a situation would require the courage to free ourselves from the constraints that prevent most of us from discovering or expressing our true nature.

Sensitive persons have always experienced a biological and emotional need for an harmonious accord with nature.

"Sometimes as I drift idly along Walden Pond," Thoreau noted in his *Journal*, "I cease to live and begin to be." By this he meant that he then achieved identification with the New England landscape.

The passive identification with nature expressed by Thoreau's phrase is congenial to Oriental thought but almost antithetical to Western civilization. Oddly enough, Tagore, a Hindu, came much closer than Thoreau to a typical Occidental attitude when he wrote that the great love adventure of European civilization had been what he called the active wooing of the earth.

"I remember how in my youth, in the course of a railway journey across Europe from Brindisi to Calais, I watched with keen delight and wonder that continent flowing with richness under the age-long attention of her chivalrous lover, western humanity. . . .

"Robinson Crusoe's island comes to my mind when I think of an institution where the first great lesson in the perfect union of man and nature, not only through love but through active communication, may be learnt unobstructed. We have to keep in mind the fact that love and action are the only media through which perfect knowledge can be obtained."[2]

The immense and continued success among adults as well as among children of *Le Petit Prince* by the French writer Antoine de Saint Exupéry (1900–1944) also reflects a widespread desire for intimate relationships with the rest of creation.

"On ne connait que les choses que l'on apprivoise, dit le renard. Les hommes n'ont plus le temps de rien connaître. Ils achêtent des choses toutes faites chez les marchands.

Mais comme il n'existe point de marchands d'amis, les hommes n'ont plus d'amis. Si tu veux un ami, apprivoise-moi!"[3]

In the popular English translation *The Little Prince*, this passage reads as follows:

" 'One only understands the things that one tames,' said the fox. 'Men have no more time to understand anything. They buy things already made at the shops. But there is no shop anywhere where one can buy friendship, and so men have no friends any more. If you want a friend, tame me.' "[4]

The French verb *apprivoiser* as used by Saint Exupéry is not adequately rendered by "tame." *Apprivoiser* implies here, not mastery of one participant over the other, but rather a shared experience of understanding and appreciation.

Poetical statements do not suffice to create conditions in which man no longer feels alienated from nature and from other men. But they are important nevertheless, because literary expressions often precede or at least sharpen social awareness. Poets, novelists, and artists commonly anticipate what is to be achieved one or two generations later by technological and social means. The poet is the conscience of humanity and at his best he carries high the torch illuminating the way to a more significant life.

Tagore wrote of man's active wooing of the earth, and stated that "love and action are the only media through which perfect knowledge can be obtained." Saint Exupéry urged that we can know and enjoy only that which we tame through love. Both have thus propounded a philosophical basis for conservation policies.

From a sense of guilt at seeing man-made ugliness, and also for reasons that must reach deep into man's origins, most people believe that Nature should be preserved. The exact meaning of this belief, however, has not been defined. There is much knowhow concerning conservation practices but little understanding of what should be conserved and why.

Conservation certainly implies a balance among multiple components of Nature. This is a doctrine difficult to reconcile with Western civilization, built as it is on the Faustian concept that man should recognize no limit to his power. Faustian man finds satisfaction in the mastery of the external world and in the endless pursuit of the unattainable. No chance for a stable equilibrium here.

To be compatible with the spirit of Western culture, conservation cannot be exclusively or even primarily concerned with saving man-made artifacts or parts of the natural world for the sake of preserving isolated specimens of beauty here and there. Its goal should be the maintenance of conditions under which man can develop his highest potentialities. Balance involves man's relating to his total environment. Conservation therefore implies a creative interplay between man and animals, plants, and other aspects of Nature, as well as between man and his fellows. The total environment, including the remains of the past, acquires human significance only when harmoniously incorporated into the elements of man's life.

The confusion over the meaning of the word Nature compounds the difficulty of formulating a philosophical basis for conservation. If we mean by Nature the world as it would exist in the absence of man, then very little of it survives. Not even the strictest conservation policies would restore the primeval environment, nor would this be necessarily desirable or even meaningful if it could be done.

Nature is never static. Men alter it continuously and so do animals. In fact, men have long recognized that they play a creative role in shaping Nature. In his *Concerning the Nature of the Gods,* written during the last century of the pre-Christian era, Cicero boasted: "We are absolute masters of what the earth produces. We enjoy the mountains and the plains. The rivers are ours. We sow the seed and plant the trees. We fertilize the earth. . . . We stop, direct, and turn the rivers; in short, by our hands we endeavor by our various operations in this world to make it as it were another Nature."[5]

For animals as well as for men, the kind of environment which is most satisfactory is one that they have shaped to fit their needs. More exactly, the ideal conditions imply a complementary cybernetic relationship between a particular environment and a particular living thing. From man's point of view, civilized Nature should be regarded not as an object to be preserved unchanged, not as one to be dominated and exploited, but rather as a kind of garden to be developed according to its own potentialities, in which human beings become what they want to be according to their own genius. Ideally, man and Nature should be joined in a nonrepressive and creative functioning order.

Nature can be tamed without being destroyed. Unfortunately, taming has come to imply subjugating animals and Nature to such an extent as to render them spiritless. Men tamed in this manner lose their real essence in the process of taming. Taming demands the establishment of a relationship that does not deprive the tamed organism—man, animal, or nature—of the individuality that is the *sine qua non* of survival. When used in the sense of the French *apprivoiser,* taming is compatible with the spirit of conservation.

There are two kinds of satisfactory landscape. One is Nature undisturbed by human intervention. We shall have less and less of this as the world population increases. We must make a strenuous effort to preserve what we can of primeval Nature, lest we lose the opportunity to re-establish contact now and then with our biological origins. A sense of continuity with the past and with the rest of creation is a form of religious experience essential to sanity.

The other kind of satisfactory landscape is one created by human toil, in which, through progressive adjustments based on feeling and thought, as well as on trial and error, man has achieved a kind of harmony between himself and natural forces. What we long for is rarely Nature in the raw; more often it is a landscape suited to human limitations and shaped by the efforts and aspirations that have created civilized life. The charm of New England or of the Pennsylvania Dutch countryside is not a product of chance, nor did it result from man's "conquest" of nature. Rather it is the expression of a subtle process through which the natural environment was humanized in accordance with its own individual genius. This constitutes the wooing or the taming of nature as defined by Tagore and Saint Exupéry.

Among people of Western civilization, the English are commonly regarded as having a highly developed appreciation of Nature. But in fact, the English landscape at its best is so polished and humanized that it might be regarded as a vast ornamental farm or park. River banks and roadsides are trimmed and grass-verged; trees do not obscure the view but seem to be within the horizon; foregrounds contrast with middle distances and backgrounds. The parklands with their clumps of trees on shaven lawns, their streams and stretches of ornamental waters achieve a formula of scenery designed for visual pleasure in the spirit of the natural conditions.[6]

The highland zone of western Britain constitutes a vast and remote area, not yet occupied by factories and settlements, offering open space for enjoyment and relative solitude. Conservation groups are struggling to protect its moors not only from industry and farming but also from reforestation. Yet the moors which are now almost treeless were once covered with an abundant growth of forest. The replacement of trees by heath and moor was not a "natural" event but one caused by the continued activities of man and his domesticated animals. Deforestation probably began as far back as the Bronze Age; the process was accelerated during the Middle Ages by the Cistercian monks and their flocks of sheep; then the exploitation of mines took a large toll of trees needed for smelting fires. In brief, the pristine ecological systems of the oak forest that once covered the highlands were eliminated by human action, leaving as relics only a few herd of deer. For nineteenth- and twentieth-century man, highland nature means sheepwalk, the hill peat bog, and the grouse moor. But this landscape is not necessarily the natural and right landscape, only a familiar one.[7]

Public attitude toward the moors is now conditioned by literary associations. This type of landscape, which exists in other parts of England, evokes *Wuthering Heights* and the Brontë sisters. Since the wild moors are identified with passionate and romantic human traits, to reforest the highlands seems to show disrespect for an essential element of English literary tradition. Similarly, the garrets of Paris, sordid as they are physically, are associated with bohemian life, Mimi in *La Bohème*, and the tunes and romance of *Sous les*

Toits de Paris. Art and literature have become significant factors in the landscape ecology of the civilized world.

The effects of history on nature are as deeply formative and as lasting as are those of early influences on individual persons and human societies. Much of what we regard today as the natural environment in England was in reality modified by the school of landscape painting in the seventeenth century. Under the guidance of landscape architects, a literary and artistic formula of naturalism transformed many of the great estates and then brought about secondarily similar modifications in large sectors of the English countryside and even of the cities. The effects of esthetic perceptions that first existed in the minds of the seventeenth-century painters thus became incorporated into the English landscape and will certainly long persist, irrespective of social changes. Less fortunately, the future development of American cities is bound to be oriented and constrained by the gridiron pattern and the network of highways which have shaped their early growth.

Profound transformations of nature by human activities have occurred during historical times over most of the world. Such changes are not all necessarily desirable, but the criteria of desirability are poorly defined. Since nature as it exists now is largely a creation of man, and in turn shapes him and his societies, its quality must be evaluated in terms not of primeval wilderness but of its relation to civilized life.

In his illuminating book *The Machine in the Garden,* Leo Marx has richly illustrated the contradictory attitudes toward Nature that have characterized American culture from its very beginning.[8] The eighteenth-century Europeans saw America as a kind of utopian garden in which they could vicariously place their dreams of abundance, leisure, freedom, and harmony of existence. In contrast, most nineteenth-century immigrants regarded the forests, the plains, and the mountains as a hideous wilderness to be conquered by the exercise of power and harnessed for the creation of material wealth.

For most people all over the world today, the American landscape still has a grandeur and an ugliness uniquely its own. Above and beyond their geologic interest and intrinsic beauty, the Rocky Mountains and the Grand Canyon of the Colorado, for example, have acquired in world consciousness a cultural significance even greater than that of the highland moors or of the Mediterranean Riviera. The beauty of America is in those parts of the land that have not yet been spoiled because they have not been found useful for economic exploitation. The ugliness of America is in practically all its urban and industrial areas.

The American landscape thus means today either the vast romantic and unspoiled wilderness, or billboards and neon signs among dump heaps. Urban and industrial ugliness is the price that America and other technological societies seem to be willing to pay for the creation of material wealth. From the wilderness to the dump appears at present to symbolize the course of technological civilization. But this need not be so, or at least we must act in the faith that technological civilization does not necessarily imply the raping of nature. Just as the primeval European wilderness progressively evolved into a humanized creation through the continuous wooing of the earth by peasants, monks, and princes, so we must hope that the present technological wilderness will be converted into a new kind of urbanized and industrialized nature worthy of being called civilized. Our material wealth will not be worth having if we do not learn to integrate the machine, the city, and the garden.

The English archaeologist Jacquetta Hawkes in *A Land* has surveyed the interplay between the people and landscape of Britain in the course of history. She presents the appealing thesis that some two hundred years ago England had come close to achieving a harmonious equilibrium between local industrial activities, the towns and villages, the farming country, and the wilderness.

"Recalling in tranquillity the slow possession of Britain by its people, I cannot resist the conclusion that the relationship reached its greatest intimacy, its most sensitive pitch, about two hundred years ago. By the middle of the eighteenth century, men had triumphed, the land was theirs, but had not yet been subjected and outraged. Wildness had been pushed back to the mountains, where now for the first time it could safely be admired. Communications were good enough to bind the country in a unity lacking since it was a Roman province, but were not yet so easy as to have destroyed locality and the natural freedom of the individual that remoteness freely gives. Rich men and poor men knew how to use the stuff of their countryside to raise comely buildings and to group them with instinctive grace. Town and country having grown up together to serve one another's needs now enjoyed a moment of balance."[9]

Even if this picture of eighteenth-century England does not exactly fit historical reality, it expresses ideals that could serve as goals for technological civilization. First it illustrates that conquest, or mastery of the environment, is not the only approach to planning nor is it the best. Man should instead try to collaborate with natural forces. He should insert himself into the environment in such a manner that he and his activities form an organic whole with Nature.

Jaçquetta Hawkes also reminds us that both the humanized landscape and the wilderness have a place in human life, because they satisfy two different but equally important needs of man's nature. Modern man retains from this evolutionary past some longing for the wilderness, even though civilization has given him a taste for farmland, parks, and gardens. Conservation policies must involve much more than providing amusement grounds for sightseers and weekend campers; they must be concerned with the biological and cultural aspects of the human past.

The huge urban areas of the modern world present problems far more complex than those of eighteenth-century England, because they are now the cradle and home of the largest percentage of mankind. We must therefore learn to provide in them means to satisfy the physiological and psychological needs of man, including those acquired during prehistory. No social philosophy of urbanization can be successful if it fails to take into account the fact that urban man is part of the highly integrated web that unites all forms of life. There have been many large cities in the past, but until recent times their inhabitants were able to maintain fairly frequent direct contacts with the countryside or with the sea. Historical experience, especially during the nineteenth century, shows that urban populations are apt to develop ugly tempers when completely deprived of such contacts. In our own times race riots provide further illustrations of this danger. Saving Nature in both its wild and humanized aspects is an essential part of urban planning.

While the problems of urbanization appear immensely complex, there is reason for optimism in the fact that some of the most crowded urban areas are also among the healthiest and most peaceful abodes of mankind. In several immense cities, furthermore, a high level of civilization has been maintained for over a thousand years. This is not the result of an accident. The medieval cities were carefully planned from their inception and life in them was rigidly controlled. The historical development of large and small European cities shows that planning and control are compatible with organic urban growth.[10]

Unfortunately, the words "planning" and "control" are in bad repute in the United States. At a recent conference on Man's Role in Changing the Face of the Earth one of the participants flatly stated that he regarded planners as akin to missionaries and preferred a world in which "there are a number of ways of living and loving and eating and drinking and building and planting and playing and singing and worshiping and thinking."[11] Freedom of behavior is important of course not only for its own sake but also because it is a condition of continued social and individual growth. On the other hand, social life is impossible without limitations to freedom. Furthermore, creativity must always be expressed within certain restraints. Almost everything that we hold dear implies restraints—from the form of a sonnet to the design of an early New England town, from the preservation of ancient monuments to familial and marital relations.

The wooing of the land in the farming areas of northern Europe, in the Mediterranean hill towns, or in the Pennsylvania Dutch country was achieved through man's willingness to accept topography and climate as a guide to planning and behavior. Many American cities—Los Angeles is only one of them—are the largest centers of nonrestraint in the world, and most of their problems derive precisely from a misapplied interpretation of freedom. In urban planning, as in all aspects of life, we must learn to discover and accept the restraints inherent in man's nature and in the conditions of our times. Civilizations emerge from man's creative efforts to take advantage of the limitations imposed on his freedom by his own nature and by the character of the land.

The Proper Study of Man in a World of Technology

There is no difficulty in recognizing among our acquaintances the basic temperaments—bilious, choleric, phlegmatic, or sanguine—that the Greek physicians described 2,500 years ago. Likewise, Shakespeare's personality types, and even his madmen, can still be identified in today's world. This permanency of man's psychological nature prompted George Bernard Shaw to remark that the length of time a literary work survives depends upon the extent to which it deals with manners, morals, or passions. Manners change so rapidly that any description of them is soon outmoded. Criteria of morality remain valid somewhat longer but they too eventually change. Passions, however, change little if at all, slowly if ever. The *Iliad* and the *Odyssey* describe adventures that have no relevance to our activities, but the passions of their heroes are still our own passions.

Love, loyalty, hatred, hunger, sexual desire, and other obvious passions are not the only aspects of man's emotional nature that have survived into modern times. Primitive people worshiped natural objects and engaged in close associations with animals. Most persons in sophisticated societies also feel the urge to associate with other forms of life; this need is commonly satisfied by establishing emotional bonds with pet animals or even with plants, but it can also find more exalted expressions that can evolve into a philosophy of life.

In her account of an African safari, Anne Morrow Lindbergh has movingly stated that the observation of animals in the wild enhanced her understanding of "how necessary life is to other life":

"Perhaps some of the tremendous renewal of energy one experiences in East Africa comes from being put back in one's place in the universe, as an animal alongside other animals—one of many miracles of life on earth, not the only miracle. Religion traditionally filled this function by giving us a sense of reverence before the mysterious forces around us; but the impact of science on our civilization had created the illusion that we are all-powerful and control the universe. . . .

"The return to reality—whether in a regional power failure or in the African wilderness—comes as a shock, but it is a healthy one. . . . In the blackout, many people rediscovered the strong web of human relatedness. In the African wilderness, man rediscovers his ancient and eternal kinship with nature and the animals."[12]

Even though he lived by hunting, primitive man worshiped animals.[13] In modern man also, the desire to hunt is paradoxically compatible with love of wild life. Hunting is a highly satisfying occupation for many persons because it calls into play a multiplicity of physical and mental attributes that appear to be woven in the human fabric. The word "paradise" has been claimed to have its origin in a Persian expression which signifies a hunting ground or at least a park enclosing animals. Certain aspects of the hunter's life are probably more in keeping with man's basic temperament and biological nature than urban life as presently practiced.

The growth of knowledge has naturally modified and enriched the expression of man's innate endowments. Religion, philosophy, and the social sciences sharpen his awareness of the fact that he is related to other men and to the rest of creation; the significance of human life is much increased by this widening of intellectual and emotional relationships. Literature and the arts enhance and educate instinctive perceptions, thus bringing into consciousness the harmonious interplay of structures and forces in the natural and the civilized world. The physicochemical and biological sciences provide, as it were, additional senses for mankind, thus broadening and intensifying man's contacts with many aspects of creation that he would otherwise ignore or but dimly perceive.

In principle, there is nothing new in the enlargement of human awareness we are witnessing at present. It is a process that has been going on for thousands of years. The practical difference is that the rate of change is accelerating and that improved communications make new experiences and perceptions more rapidly available to the rest of mankind. Although experiences and perceptions are inevitably distorted as they are transmitted from person to person, further and further away, the important thing is that they are not lost. Even in their distorted version they affect tastes, opinions, attitudes, and, in the end, goals. The future of human societies is entirely open because new evolutionary trends are bound to result from the emergent novelty.

The wide range of possibilities offered to human life by sociocultural evolution accounts for the fact that all utopias have been stillborn or have soon disintegrated and vanished. Static utopian societies might provide comfortable lives for correct gray men in gray flannel suits. But vital human beings would soon want to escape from Utopia in the hope of fulfilling themselves more completely in more dynamic environments.

Utopias are based on the assumption that what is known today represents most of what can be known, whereas the real world is forever changing. Furthermore, man has so many unexpressed potentialities that a new creation is likely to emerge whenever he encounters new circumstances. Civilizations remain capable of surviving and growing only if they enable man to express the aspects of his nature that are still dormant. Growth is the outcome of the interplay between the human endowment, which is essentially unchangeable but rich in unexploited resources, and the total environment, which continuously evolves through the integration of natural forces with human activities.

Modern cities, especially the American urban megalopolis, are becoming a nightmare because they increasingly fail to provide a satisfactory environment for the unchangeable requirements of man's biological nature and for his cultural evolution. They will become even worse on both these accounts if urban planners continue to be concerned chiefly with economic and technological criteria instead of directing their attention to the factors that favor a healthy and civilized life.

During the past few decades, architects and urban planners have proposed highly imaginative models of futuristic cities, extravagant either in length or in height, incorporating all existing or imaginable technological controls of space and communication, fully equipped with automated power equipment and facilities for effortless amusement. The very diversity of the models now being presented for the city of the future is evidence that a philosophy of urban life does not yet exist. Such intellectual free-wheeling is possible only because the planners appear to be unconcerned with, unless they be ignorant of, fundamental human needs. They seem to accept at its face value the statement attributed to the American technologist Buckminster Fuller that architecture is a "technical optimum per pounds of invested resources."[14] As if the really significant criteria of planning and architecture were cost and efficiency of buildings rather than the suitability of environments for human needs, potentialities, and aspirations! In his novel *Fahrenheit 451*, the American science-fiction writer Ray Bradbury illustrates vividly the consequences to be expected from highly artificial ways of life and technicized environments in which human beings are deprived of the natural stimuli under which *Homo sapiens* evolved.[15]

Secretary of the Interior Stewart Udall has repeatedly expressed his dismay at the neglect of human needs in technological engineering. As he points out: "Certain brilliant men are so engrossed in engineering techniques that they have seemingly lost sight of their own species. Buckminster Fuller, one of the most creative of our designers, has proposed that we build gargantuan geodesic domes over our cities. These great greenhouses would enclose a mechanized, man-controlled climate: the stars, the seasons, and the sun would be walled out in a triumph of technology. Air pollution *and* weather would disappear; yet these domes would deny the instinct of man to coexist with nature."[16]

A multiplicity of direct human contacts is one of the

factors requisite for the normal development of the mind and for emotional equilibrium. Even the most efficient technological methods of communication cannot replace direct encounters as a basis for a healthy, happy, and creative existence. We learn chiefly from others, and most effectively as the result of I-Thou relationships. Properly directed biological studies could show that, in men as in animals, there are sociobiological requirements that will not be satisfied if the city of the future fails to provide suitable conditions for numerous and undisturbed face-to-face relationships. The visit of the Russian premier Aleksei Kosygin to the United Nations Assembly in New York provided a spectacular illustration of the need for direct encounters. When he and President Johnson decided to meet in a small New Jersey town, it was not for the purpose of negotiating agreements or exchanging information, but only in the hope of gaining better understanding of each other. Written documents, color television, and hot-line messages between Washington and Moscow could not substitute for even short periods of face-to-face encounters.

Since environmental factors profoundly condition most aspects of daily existence, and in particular the biological and psychological development of children, the most urgent need in urban planning is a better knowledge of what human beings require biologically, what they desire culturally, and what they hope to become. In this, as in all important aspects of life, the know-how is less important than the know-why; unfortunately, technological considerations are practically always given precedence over human factors. Modern man finds it easy to function as *Homo faber*, whether he produces automobiles, highways, skyscrapers, guided missiles, or no-calorie foods. But he has not yet learned to function as *Homo sapiens* when it comes to using wisely the objects that he makes in such nauseating profusion.

Cities, dwellings, and the ways of life in them cannot be designed or imagined merely on the basis of available technology. Each decision concerning them must take into consideration not only human needs in the present but also long-range consequences. A design for living that would provide the opportunity for invigorating walks in a pleasant and unpolluted atmosphere would contribute more to physical and mental health than any concern with availability of elevators, moving platforms, mechanical hearts, and psychiatric care.

Most importantly perhaps, the city should provide a great diversity of environments so as to encourage the expression of desirable human potentialities that might otherwise remain latent. Throughout history different social groups have expressed the multifarious activities of the human mind in a great variety of architectural creations: monasteries and cathedrals, private gardens and public parks, highways and byways, lecture halls and Disneylands. Even if the designs of modern urban developments were really compatible with the biological success of *Homo sapiens*—a doubtful assumption—they would be objectionable because their monotony and dreariness do not encourage the unfolding of the most desirable human characteristic.

Our societies will inevitably experience biological and psychological disasters unless they develop technological and urban environments really suited to human needs. The diseases of civilization and the rebellion of youth are warnings that physical welfare, mental sanity, and emotional satisfaction require more than economic affluence, production of things, and knowledge of molecules.

Even if children now growing up enjoy economic prosperity in a world of peace during the forthcoming decades, they will suffer from their surroundings and revolt, or at least become disenchanted, unless we correct the technological and biological absurdities of modern life. The most spectacular achievements of the first seven decades of the twentieth century would pale in significance if we did manage to create environments in which human beings, and especially children, could safely express the rich diversity of their genetic endowment. But this will require an immediate effort to change the course of science and of technology.

The interdependence between the whole man and his total environment generates theoretical and practical problems that should be studied by scientists, especially those concerned with human biology and technology. One of the distressing oddities of the scientific era, however, is its failure to apply the methods of science to the most important problem of human life. Research institutions, in or outside universities, are equipped physically and intellectually for studying in great detail man as a machine, but they either neglect entirely or give a low order of priority to the problems that living man encounters in the course of his daily life.

When the National Academy of Sciences celebrated its centenary in 1963, some twenty illustrious American scientists presented various facets of the "scientific endeavor" in a series of brilliant essays on topics ranging from astronomy to animal behavior.[17] Not one of them, however, touched on the problems that the man of flesh and bone meets in the twentieth-century world. This omission was not the result of an oversight; it reflected the fact that the way the problems of human life in the real world are being studied does not provide knowledge of sufficient intellectual distinction for the august atmosphere of the National Academy of Sciences. Origin of the solar system or electron microscopy of dead animal tissues, yes. Man suffocating and experiencing alienation in our cities, no.

Scientists shy away from the problems posed by human life because these are not readily amenable to study by the orthodox methods of the natural sciences. For this reason, such problems are not likely to yield clear results and rapid professional advancement. The way to scientific success is often through substituting for important problems that appear overwhelmingly complex other far less important problems that can be solved within a relatively short time. This situation is not peculiar to capitalistic societies. The Soviet physicist L. A. Artsimovitch is reported to have said: "Scientific research is a method whereby private curiosity is satisfied at public expense." This phrase is not as facetious as it sounds; it points to a serious problem concerning the social relations of science.

There is a more honorable, though fallacious, reason for the reluctance of the scientific establishment to initiate, encourage, or support the study of the complex problems of human life. It is the strange assumption that knowledge of complex systems will inevitably emerge from studies of much simpler ones. Among biologists, this attitude has generated a widespread belief that natural phenomena and living organisms can best be studied by dividing them into their component parts. The obvious appeal of this reductionist approach is that elementary structures and properties thus isolated can be analyzed in greater and greater detail by the orthodox methods of physics and chemistry.

The reductionist approach has been immensely fruitful in discoveries and has made it possible to convert certain aspects of knowledge into power. Unfortunately, it has resulted in the neglect of many important fields of science and has encouraged an attitude toward Nature which is socially destructive in the long run. The study of the *interplay* between the component parts of the system is at least as important as the study of any or all of the isolated components. The destructive mismanagement of human lives and of natural resources is due more to our neglect of the interplay between the various forces operating in the modern world than to ignorance concerning these forces themselves.

The reductionist scientist tends to become so involved intellectually and emotionally in the elementary fragments of the system, and in the analytical process itself, that he commonly loses interest in the phenomena or the organisms which had been his first concern. For example, the biologist who starts with a question formulated because of its relevance to human life is tempted, and indeed expected, to progress *seriatim* to the organ or function involved, then to the single cell, then to subcellular fragments, then to molecular groupings or reactions, then to the individual molecules and atoms. He would happily proceed, if he knew how to do it, until he reached the ultimate aspects of nature in which matter and energy become indistinguishable. Problems of great interest arise at each step of the reductionist process, but in practically all cases the original problem posed by the organism or the phenomenon itself is lost on the way. Many books are being written on the theme "from molecule to man," but they have surprisingly little to say about man or the problems that really matter in human life. Scientists, like technologists, find it more comfortable to function professionally as *Homo faber* than as *Homo sapiens*.

In order to understand the mechanisms through which natural systems function in an integrated manner, the study of parts must be complemented by ecological studies of systems functioning as integrated wholes. No one would assume that, because decisions to buy or sell are mediated through the passage of nerve impulses along the neurons, the science of economics has much to gain from discoveries in molecular neurophysiology, nor is there any illusion that the understanding of digestive processes will be advanced by studies of quantum mechanics. Yet it is often claimed that problems of human life cannot be studied scientifically until more has been learned of the submicroscopic particles and enzymatic processes involved in cellular functions. This is not only intellectual nonsense; it is a form of escape from social responsibility.

Each particular field of science has its own logic of growth and must develop its own techniques. Many of the problems that mankind faces today are the consequences of disjunction between man's nature, his environment, and the creations of scientific technology.

Most of man's problems in the modern world arise from the constant and unavoidable exposure to the stimuli of urban and industrial civilization, the varied aspects of environmental pollution, the physiological disturbances associated with sudden changes in ways of life, the estrangement from the conditions and natural cycles under which human evolution took place, the emotional trauma and the paradoxical solitude in congested cities, the monotony, boredom, and compulsory leisure—in brief all the environmental conditions that undisciplined technology creates. These influences affect all human beings in affluent countries, irrespective of genetic constitution. They are not inherent in man's nature but are the products of his responses to social and technological innovations.

Since all aspects of human life reflect environmental influences, it is a moral obligation for the scientific community to devote itself in earnest to the study of ecological systems, both those of nature and those created by man. A new kind of knowledge is needed to unravel the nature of the cohesive forces that maintain man in an integrated state, physically, psychologically, and socially, and enable him to relate successfully to his surroundings. Hardly anything is known of his adaptive potentialities, of the manner in which he responds to the stimuli which impinge on him early in his development and throughout his life, of the long-range consequences of these responses not only for himself but for his descendants. These and countless other problems of human life should and could be studied scientifically yet have hardly any place in the curriculums of universities or research institutes.

Man's responses to his total environment can be brought within the scope of science by developing laboratory models to supplement observations on man. Experimental studies might deal at first with the effects on health, behavior, and performance of environmental factors such as shape and size of rooms, air-conditioning, kinds of intensity of stimulation, crowding, transient isolation, or any influence that can be manipulated and to which man is likely to respond, consciously or unconsciously. As work proceeds along such lines, new techniques and new scientific goals will certainly emerge and thus progressively lead to the development of a true science of human life in the modern world.

The more life becomes dependent on technology, the more it will be vulnerable to the slightest miscarriage or unforeseen consequence of innovations; hence the need for studies of interrelationships within complex systems. Science will remain an effective method for acquiring knowledge meaningful to man only if its orthodox techniques can be supplemented by others which come closer to the human experience of reality. To serve human welfare, action must be guided by a better knowledge of fundamental human needs. A truly human concept of technology might well constitute the force that will make science once more part of the universal human discourse, because technology at its highest level should integrate the external world and man's nature.

Introducing science into human life enlarges the scope of freedom and responsibility. In most cases, choices and decisions have to be made on the basis of value judgments which transcend knowledge and involve not only the here and now but also anticipations of the future. Scientific understanding helps in predicting the likely consequences of social and technological practices; it provides a more rational basis for option. Since awareness of consequences usually plays a part in decision-making, scientific knowledge could become one of the criteria for the acceptance or rejection of old value systems, and even for the development of new ones.

The management of natural forces through technology has been so far the most characteristic urge as well as the most spectacular achievement of Western civilization; science owes its popular prestige to its technological applications rather than to its conceptual content. Despite its successes in practical fields, scientific knowledge *per se* is not yet established as a genuine and meaningful value in the public mind. As stated by the English historian of science Stephen Toulmin, "Western society today may be said to harbour science like a foreign god, powerful and mysterious. Our lives are changed by its handiwork, but the population of the West is as far from understanding the nature of this strange power as a remote peasant of the Middle Ages may have been from understanding the theology of Thomas Aquinas."[18]

The scientists' neglect of the problems which are of deepest concern to humanity could well transform the antisocial and anti-intellectual outbursts of the present period into an anti-science crusade. Educators and sociologists have been alarmed by a trend away from the natural sciences among young people in Great Britain and the United States, despite the great inducements and public pressure to encourage and facilitate the development of scientific careers. There are probably many reasons for this attitude among young scholars; one may be that in their present form the natural sciences and the technologies derived from them do not satisfy the deep social concern so prevalent among the modern generation. Students are beginning to doubt that Galileo, Watt, and Edison have contributed as much and as lastingly to human advancement and happiness as Socrates, Lao-tze, and Francis of Assisi.

Everywhere in the Western world, youth is protesting against the overorganization and dehumanization of society. They side with contemporary American writers such as Henry Miller, who brings esthetic eroticism into literature, and Norman O. Brown,[19] who advocates the resurrection of the body, because they are more receptive to relevant irrationality than to irrelevant rationality. These attitudes should not be dismissed as trivial and irresponsible. They are the manifestation of vital forces that are essential for the sanity of mankind and that must be satisfied in order to assure continued creativity.

In science as in other human activities, the speed of progress is less important than its direction. Ideally, knowledge should serve understanding, freedom, and happiness rather than power, regimentation, and technological development for the sake of economic growth. Emphasis on humanistic criteria does not imply a retreat from science; rather it points to the need for an enlargement and a rededication of the scientific enterprise. Scientists must give greater prominence to large human concerns when choosing their problems and formulating their results. In addition to the science of things, they must create a science of humanity, if they want the intellectual implications and practical applications of their efforts to be successfully woven into the fabric of modern life. Cultures and societies, just like other living organisms, cannot survive if they do not maintain internal integration. Science can become fully integrated in the sociocultural body only if it achieves a more meaningful relationship to the living experience of man.

Man Makes Himself

The most ancient written documents that have come down to us make clear that mankind was already beset with problems of social ethics at the beginning of its history in Mesopotamia. One of the oldest clay tablets from Sumer, some 5,000 years old, carries a message from the king forbidding the high priest to go "into the garden of a poor mother and take wood therefrom, or gather tax in fruit therefrom." Hammurabi who reigned in Babylon around 2000 B.C. claimed that the Lord of Heaven and Earth had given him the task "to prevent the strong from oppressing the weak; . . . to enlighten the land and to further the welfare of the people."

In the Sumerian language, the word *namlulu* was used at first to denote mankind, or human beings in the collective sense. Its meaning evolved with time and in later texts it came to refer to the behavior and attributes of mankind. In other words, *namlulu* eventually acquired an abstract sense somewhat similar to that of the English word "humanity." This is illustrated by a text on a Sumer clay tablet in which a father upbraids his son for ingratitude and laziness, also for failing to follow in his own footsteps and become a scribe: "Night and day you waste in pleasure. You have accumulated much wealth, have expanded far and wide, have become fat, big, broad, powerful, and puffed up. But . . . you looked not to your humanity."[20]

It is impossible to retrace from inscriptions on stone and clay tablets the many nuances through which the original meaning of the Sumerian word *namlulu* progressively became transformed into one symbolizing the abstract and complex values now associated with the word humanity. From a general point of view, however, the historical evolution of most important concepts has consisted in the progressive replacement of subconscious and half-conscious mental perceptions by increasingly conscious analytical thinking.

In the past, even the most sophisticated people commonly employed a fictional form to illustrate and indeed to convey their messages. This is true of Plato, and of course even more of Jesus. Factual knowledge and conceptual truth have reached us from ancient times largely in the form of parables. The word "parable" is of Greek origin and means "to throw across"; it refers to a story that bridges the imagined and the real. For the past three thousand years we have been engaged in the painful and often dull process of converting these parables, the picturesque and brilliant imaginings of our ancestors, into objective reality. Ancient writings described man's behavior in images that are still meaningful for us; modern physiologists and psychologists now try to discover the mechanisms of these intuitive perceptions.

When Aristotle wrote in the fourth century B.C. that the "nature of man is not what he was born as, but what he is born for," he tacitly implied some relationship between man's biological origin and the ideals of human destiny. Modern scholars now try to comprehend how the biological endowments of the species *Homo sapiens* have progressively generated the spiritual values that give to human history a number of characteristics not found in animal life. Although *Homo sapiens* has remained much the same biologically since Paleolithic times; human life has continued to evolve through sociocultural mechanisms. We are increasingly conscious of this change and try deliberately, even though clumsily, to govern our lives more rationally. Whereas Aristotle believed *a priori* that we are "born for" some kind of purpose, we would prefer to decide for ourselves what this purpose should be.

It is unlikely that Paul the Apostle had anything like the modern theory of evolution in mind when he wrote of human nature: "It is sown a natural body; it is raised a spiritual body. . . . The first man is of the earth, earthy: the second man is the Lord from Heaven" (I Corinthians 15: 44,47). Yet the Apostle's statement does express symbolically the belief that man has escaped from the bondage of his past and is creating ways of life in which animal instincts and urges are progressively modified and supplemented by intellectual preoccupations and spiritual values.

Human beings experience the world through their senses, but paradoxically what they prize most is often independent of sense perceptions. Indeed, many have sacrificed their biological existences at the altar of nonmaterial values conceived in the soul rather than experienced in the flesh. The biblical injunction that he who would save his life must first be willing to lose it may seem obscure to us, but it has proved deeply meaningful to the many men in all times who have used it as an ethical guide. Prophets and religious leaders have preached the biblical injunction as a revealed truth; philosophers acknowledge its human significance and try to discover its precise historical, psychological, and ethical meaning.

Each type of civilization has contributed its share to the conscious analytical definition of ancient intuitive wisdom and to man's confidence that he can shape his destiny. The writings of the fifteenth-century Italian humanist Conte Giovanni Pico della Mirandola constitute a landmark of particular interest for our own civilization because they express the spirit of the Italian Renaissance at the dawn of the scientific era. In his *Oration on the Dignity of Man*, Pico rejected what was then the traditional view that man owned his uniqueness to his place in the center of things and to inherent qualities that differentiated him from animals. He asserted instead that man has no fixed properties but has been endowed by God with the power and the responsibility for choosing the forms and values of his life:

"The Great Artisan created man with an undetermined nature, and then told him: you . . . shall determine for yourself your own nature, in accordance with your own free will. . . . Neither heavenly nor earthly in nature, you may fashion yourself in whatever form you shall prefer. You shall have the power to degenerate into the lower forms of life, which are brutish. But you shall also have the power, out of your soul's judgment, to be reborn into the higher forms, which are divine."[21]

The Renaissance scholars derived from Pico's advocacy of man's freedom and ability to develop his innate nature a creative and all-encompassing doctrine of the humanities. The belief that human life is a continous self-creation has continued to dominate scholarship into our times and has been forcefully expressed in the phrase "Man makes himself" which Gordon Childe used as the title for his classic book on social anthropology. The last sentences of that book constitute the modern scientific equivalent of Pico della Mirandola's assertion: ". . . because tradition is created by societies of men and transmitted in distinctively human and rational ways, it is not fixed and immutable; it is constantly changing as society deals with ever new circumstances. Tradition makes the man, by circumscribing his behaviour within certain bounds; but it is equally true that man makes the traditions. And so, we can repeat with deeper insight, 'Man makes himself.' "[22]

For almost a century, proponents of eugenics have preached that the human species could be upgraded by judicious breeding policies and especially by eliminating the unfit from the reproductive pool. A few illustrious geneticists have recently reformulated the eugenics creed in terms of modern genetic theory. In particular, they have advocated selective reproduction through the use of sperm from donors having desirable qualities. Other very sophisticated methods of controlled genetic reproduction are being discussed at the present time.[23] None of these approaches to improvement of the human stock will be considered here, because there is as yet no scientific basis for justifying any attempt at modifying, let alone improving, man's genetic make-up. In view of our abysmal ignorance of the really important aspects of human genetics, it is just as well that methods of genetic control would not be socially acceptable.

In practice, the control and improvement of human life has so far been achieved, and will continue to be achieved, by manipulating the social and physical factors of the environment. There is only one way to cope with and take advantage of man's genetic diversity; it is to diversify man's environment. Although this has long been empirically recognized and practiced, a systematic effort to manipulate the interplay between man and his environment through sociopolitical measures did not consciously begin until the eighteenth century. From then on, sociologists have acted on the belief that all aspects of human life are affected by the total environment and therefore can be manipulated socially.

Concern for the quality of the environment achieved a rational and coherent expression during the second half of the nineteenth century. In Western Europe and then in the United States, the early phases of the Industrial Revolution had resulted in crowding, misery, accumulation of filth, horrible working and living conditions, ugliness in all the mushrooming industrial areas, and high rates of sickness and mortality everywhere. The physical and mental decadence of the working classes became intolerable to the social conscience and in addition constituted a threat to the future of industrial civilization.

The social response took many forms, one of the most original and effective being a systematic effort to improve the quality of life by correcting the physical environment. "Pure water, pure air, pure food" was the motto around which the campaign for social and environmental reforms was initially organized. Efforts were also made at that time to improve housing conditions and to reintroduce in rural and especially in urban life some of the amenities and values that had been destroyed by industrialization. Country lanes and waterways, boulevards adorned with trees and flowers loomed almost as large as sanitation in the plans of nineteenth-century reformers.

Two books, both published in the 1870s, stand out as expressing in the form of practical instructions the philosophy of environment that then prevailed in the Western world. These are *The Value of Health to a City* by the Munich sanitarian Max von Pettenkoffer (1873), and *Hygeia: A City of Health* by the Englishman B. W. Richardson (1876).[24] The programs for urban planning and the specifications for housing formulated by Pettenkoffer and Richardson are typical of those that proved influential in improving the health of city dwellers in the countries of Western civilizations around the turn of the century. Yet the message of their books is now of historical interest only. Today's environmental problems are different and far more intractable. The exhaustion of natural resources and the erosion of the land; the chemical pollution of air and water; the high levels of noise, light, and other stimuli; the pervasive ugliness and inescapable pressures resulting from high-population densities and mechanized life; all these phenomena and many others which threaten the life of modern man have become critical only during recent decades.

Our nineteenth-century forebears approached their problems through a creative philosophy of man in his environment. In contrast, we tend to act only under the pressure of emergency. During recent decades, for example, the construction of great dams all over the world has been prompted not by a comprehensive, integrated program of land-and-water use but by the threat of destructive floods and by shortages of safe water supplies. It took the tragedies of the dust bowls in the 1920s and 1930s to activate policies for the control of soil erosion. Communities are only now waking up to the dangers—which could have been predicted decades ago—arising from undisciplined technology and population growth. The monstrous ugliness of our cities and highways is generating some concern for their esthetic improvement. The scandal of the living conditions in the slums of large cities entered public consciousness only under the pressure of race riots. However, awareness of these problems does not seem sufficient to generate really effective control policies.

What is called "environmental improvement" merely consists in most cases of palliative measures designed to retard or minimize the depletion of natural resources, the rape of nature, the loss of human values, and social unrest. Such programs can be regarded at best as short-sighted adaptive responses to acute crises. They are the expressions of fear or panic rather than of constructive thought.

It is only fair to mention that a few legislators have emphasized that the government has fallen behind in a difficult game by reacting to crises instead of anticipating them. For example, Senator E. S. Muskie and Representative E. Q. Daddario are presently attempting to take an overall view of technology and of its impact on the environment and on man's welfare; they have formulated legislative programs to deal with these problems. While I admire the wisdom and courage of these legislators, I am afraid that their well-conceived efforts will not be fruitful unless an environmental catastrophe frightens the public sufficiently to force Congress into action.

Despite the intensity of social and racial conflicts everywhere in the world, there is reason to believe that progress is being made toward the solution of the political problems which now plague mankind. It is almost universally accepted, in principle at least, that all human beings are entitled to freedom, social justice, and equal opportunity; many collective and individual efforts are now being made to implement these concepts. In contrast, there is as yet no social philosophy of what should be done to improve urban and rural environments. Programs for the Great and Beautiful Society are a hodge-podge collection of measures hastily formulated to correct a few glaring defects in cities and in the countryside. They are a far cry from a social philosophy for integrating urban, rural, and wilderness areas in such a manner as to provide the modern equivalent of the eighteenth-century England described by Jacquetta Hawkes.

Contrary to what is generally claimed, increased knowledge of natural forces and the growth of technology have not improved man's control over the environment. While the rate of environmental change has immensely accelerated, the social and biological responses have not kept pace with the new situations thus created. As a result, technicized societies may be close to the threshold beyond which it will be impossible to evaluate, let alone control, the effects on human life of the new environments created by technological innovations.

In *The Technological Society*, Jacques Ellul has asserted that the technical take-over of life has already begun and has gone so far that it may be irreversible. The fact that Ellul's book has called forth such widespread and intense reactions, both favorable and hostile, indicates the depth of the public concern over the possibility that technology has indeed "become an end-in-itself, to which men must adapt themselves."[25]

The problems posed by the social control of technology are much the same under capitalism, socialism, or communism. Irrespective of political philosophy, new formulas of social planning must be discovered to make technology subservient to worthwhile human needs, instead of allowing it to grow for its own sake or as a tool for economic or national expansion. It is tragically symbolic of the distorted sociotechnological philosophy which now governs human life that immense efforts are being made to develop new coating and finishing techniques to protect automobile bodies against the corroding effects of air pollutants, whereas hardly anything is being done to study the effect of pollution on the human body!

Many sociologists who take for granted the need of Keynesian economic controls advocate a laissez-faire attitude toward technology. Their argument is much the same as that once used to defend laissez-faire in economics—that control stifles initiative and discourages innovations. Yet planning for better-defined and worthwhile human goals has become urgent if we are to avoid the technological take-over and make technology once more the servant of man instead of his master.

The phrase social planning is commonly associated with political utopias and for this reason elicits skepticism and even hostile reactions, especially in the United States. Utopias are no longer fashionable, in part as a result of the progressive erosion of the belief in rational progress, and more justifiably because of the awareness that static institutions cannot survive in a competitive world. The regrettable consequence of this skepticism, however, is that intellectuals and social critics have tended to be satisfied with ridiculing the times in which we live and with describing anti-utopias instead of utopias. Yet to formulate constructive alternatives for existing institutions is more important, though more difficult, than to caricature the present state of affairs or simply protest against social evils.

Since an immense amount of money and effort will certainly be expended on programs of social and environmental improvement in the near future, it is essential that we try to imagine the kind of world we want. All great periods of history have created such utopian images. Inertia is the only mortal danger. Like the poet, the social planners should break for us the bonds of habit. In the words of Oscar Wilde: "A map of the world that does not include Utopia is not worth even glancing at, for it leaves out the one country at which Humanity is always landing. And when Humanity lands there, it looks out, and seeing a better country, sets sail. Progress is the realization of Utopias."[26]

Most human problems have such complex historical and social determinants that they do not lend themselves readily to tidy planning or to study by the methods of the natural sciences. Their complexity comes from the fact that they involve not only man's biological needs in the here and now, but also his past, his potentialities, and his limitations. Planning for urban and rural development, as well as the design of public buildings or private residences, will remain empirical activities until based on better knowledge of the long-range effects that environmental factors exert on physical and mental health. The American authority on urban problems Charles Abrams has expressed the same thought: "In interviewing an architect in 1948 who was then planning the Cleveland zoo, I was struck by the quantity of research that goes into the study of animal habits. The general instructions to the architects were to retain the natural values of sites, simulate the natural habitats of each animal, and guarantee freedom from unnecessary distractions as well as absolute privacy for copulation. Specialists from all over the world were consulted on the eating, sleeping, and mating habits of each species, and the findings were reduced to detailed reports which were carefully studied before a line was drawn. No comparable studies, to my knowledge, have ever been made on the human animal in its urban surroundings nor are we even as much concerned as the zoo architect about the human habitation. The sciences of urban anthropology and human nidology, particularly as they bear on the human female, are not even at their beginnings."[27]

Only a few generalizations can be offered here to illustrate the extent of our ignorance concerning man's responses to environmental and social forces.

Everyone agrees that all citizens should be given equal educational opportunities. But what are the critical ages for the development of mental potentialities and for receptivity to the various kinds of stimuli? What, in this connection, are the effects of prenatal and early postnatal influences on the physical, physiological, and mental characteristics of the adult? Which of these effects are irreversible? To what extent and how can the effects of early deprivations be corrected?

Everyone agrees that our cities must be renovated, or even rebuilt. But while technologies are available for almost any kind of scheme imagined by city planners, architects, and sociologists, who knows enough to tell, or who tries to discover, how the environments so created will affect human well-being and condition the physical and mental development of children?

Everyone agrees that it is desirable to control environmental pollution. But which pollutants of air, water, or food are really significant? The acute effects of pollutants and drugs can be readily recognized, but what about the cumulative, delayed, and indirect effects? Does the young organism respond as does the adult? Does he develop forms of tolerance or hypersusceptibility that affect his subsequent responses? Without such knowledge, priorities in the control of environmental pollution or of drug usage cannot be established rationally.

These few examples will suffice to illustrate that the environment is being studied almost exclusively from a technological point of view without much concern for its biological and psychological effects.

John Donne made us conscious of the fact that no man is an island and that the bell tolls for all of us, but it was the last speech delivered by Ambassador Adlai E. Stevenson that made man's dependence not only on other men, but also on the resources of the earth, one of the most poignant issues of our times: "We all travel together, passengers on a little space ship, dependent on its vulnerable supplies of air and soil; all committed for our safety to its security and peace, preserved from annihilation only by the care, the work, and I will say the love we give our fragile craft."[28] Writing at the turn of the seventeenth century, Donne used the image "island" to convey man's dependence on his neighbors, but so contemporary a person as Stevenson changed the parable and identified the human condition with life on a spaceship, in which all aspects of creation are interdependent.

Before long, all parts of the globe will have been colonized and the supply of many natural resources will have become critical. Careful husbandry, rather than exploitation, will then be the key to survival. Developing stations in outer space or on the bottom of oceans will not modify significantly, if at all, the limitations of human life. Man emerged on the earth, evolved under its influence, was shaped by it, and biologically he is bound to it forever. He may dream of stars and engage in casual flirtations with other worlds, but he will remain wedded to the earth, his sole source of sustenance.

As the world population increases, the topographical limitations of the spaceship Earth and the exhaustion of some of its natural resources will inevitably require that its economy be based on strict ecological principles. This imperative necessity, however, is not yet widely recognized. The very word ecology was introduced into the scientific language only seventy-five years ago—so recent is the awareness that all components of nature are interwoven in a single pattern and that we too are part of the pattern.

Until now, man has behaved as if the areas available to him were unlimited, with infinite reservoirs of air, soil, water, and other resources; he could do this with relative impunity in the past because he could always find some other place in which to start a new life or engage in any kind of adventure that he chose. There was always a new Jerusalem beyond the hill. Since the evolutionary and historical experiences of man are woven in his mental fabric he naturally finds it difficult to rest quietly in a corner of the earth and husband it carefully. His thoughtlessness in provoking situations that are potentially dangerous arises partly from the fact that he has not yet learned to live within the constraints of the spaceship.

The ecological attitude is so unfamiliar, even to many scientists, that it is often taken to imply acceptance of a completely static system. Students of sociology have expressed concern less the ecologists' delight in the well-balanced, smoothly functioning, steady-state ecosystem of the pond be extrapolated uncritically to the whole earth and its human population. If the ecologists' concept of man's relation to the total environment really did imply a steady-state system, ecological philosophy would indeed be dangerous as well as wrong, because it would imply that the human adventure has come to an end. But this need not be the case.

The physical forces of the environment are forever changing, slowly, but inexorably. Furthermore, all forms of life including human life are continuously evolving and thereby making their own contributions to environmental changes. Since man's nature leads him to search endlessly for new environments, and for new adventures, there is no possibility of maintaining a *status quo*. Even if we had enough learning and wisdom to achieve at any given time an harmonious state of ecological equilibrium between mankind and the other inhabitants and components of the spaceship Earth, it would be a dynamic equilibrium, which would be compatible with man's continuing development. The question is whether the interplay between man and his natural and social surroundings will be controlled by blind forces, or whether it can be guided by deliberate, rational judgment.

Admittedly, all of human evolution and most of human history have been the result of accidents or blind choices. Many deliberate actions have had unforeseen consequences that proved unfortunate; in fact, most of the environmental problems that now plague Western civilization derive from discoveries and decisions made to solve other problems and to enlarge human life. The internal-combustion engine, synthetic detergents, medicinal drugs, and pesticides were introduced with useful purposes in mind, but some of their side effects have been calamitous.[29] Efficient methods of printing have made good books available at low prices but are now cluttering mailboxes with despicable publications and useless advertisements and burdening waste baskets with mountains of refuse that must be burned and thus pollute the air.

We may hope eventually to develop techniques for predicting or recognizing early the objectionable consequences of social and technological innovations so as to minimize their effects, but this kind of piecemeal social engineering will be no substitute for a philosophy of the whole environment, formulated in the light of human aspirations and needs. We cannot long continue the present trend of correcting minor inconveniences and adding trivial comforts to life at the cost of increasing the likelihood of disasters and cheapening the quality of the living experience. If the goal of technological civilization is merely to do more and more of the same bigger and faster, tomorrow will only be a horrendous extension of today.

Creating a desirable future demands more than foresight; it requires vision. Like animal life, human life is affected by evolutionary forces that blindly shape the organism as it responds to its environment. Human history, however, involves also the unfolding of visionary imaginings. The philosophers of the Enlightenment had imagined the modern world long before there was any factual basis for their vision. They prepared the blueprint for most of the desirable aspects of modern life in the faith that objective knowledge, social reforms, and scientific technology could someday liberate human beings from fear and destitution. Throughout human history, progress has thus been a movement toward imagined goals; the realization of these aims has in turn inspired new goals.

Mankind's greatest achievements are the products of vision. This statement could readily be illustrated with examples taken from scientific history, but phenomena of ordinary life will serve just as well. One needs only think of the marvelous parks and gardens of Europe to realize the creative force of a long-range view in social improvement. Parks and gardens originated from that extraordinary sense which is peculiar to man, the vision of things to come. Books by the landscape architects of the eighteenth century contain drawings of these parks as they appeared at the time of their creation, with the naked banks of newly created brooks and lakes among puny trees and shrubs—landscapes without substance or atmosphere. Yet the landscape architects had composed the expanses of water, lawns, and flowers to fit the silhouettes of trees and the masses of shrubbery, not as these components of the scenery existed when first put together, but as they were to become with the passage of time. The architects had visualized the future and then drew plans to make it come true.

In *The Design of Cities,* the American city planner Edmund Bacon has described with obvious admiration the progressive development over several centuries of some of the great urban sites and vistas of Europe. For example, the Piazza del Popolo in Rome and the Tuileries-Champs Elysées-Etoile complex in Paris had been visualized long before social conditions and economic resources justified their existence or made them possible.[30]

While the great European gardens, parks, and urban vistas still delight us today, other kinds of landscapes must be conceived to meet present and future needs. The old country roads, lined with stately trees, provided poetic and practical shelter for the man on foot or horseback and for coaches; a modern highway, however, must be designed in such a manner that horizons, curves, and objects of view are related to the physiological needs and limitations of motorists moving at high speed. The evolution from park to parkways involves biologic and esthetic factors as much as technologic determinants.

We would do well to keep in mind the advice given a century ago by the American landscape architect Frederick Law Olmsted, who designed Central Park in New York and several other wonderful parks in American cities. "In the highways, celerity will be of less importance than comfort and convenience of movement; and as the ordinary directness of line in town-streets, with its resultant regularity of plan would suggest eagerness to press forward, without looking to the right hand or to the left, we should recommend the general adoption, in the design of your roads, of gracefully-curved lines, generous spaces, and the absence of sharp corners, the idea being to suggest and imply leisure, contemplativeness and happy tranquillity."[31]

Ray Bradbury's vision of automobile travel in the future world as symbolized in *Fahrenheit 451* provides a vivid contrast to Olmsted's wise advice. " 'Have you seen the two-hundred-foot-long billboards in the country beyond town? Did you know that once billboards were only twenty feet long? But cars started rushing by so quickly they had to stretch the advertising out so it would last.' "[32]

Envisioning an environment suitable for the total life of an immense technological society is vastly more complicated than visualizing the future appearance of a park or designing a parkway. But certain principles hold true for all environmental planning, because they are based on unchangeable aspects of man's nature.

On the one hand, the genetic endowment of *Homo sapiens* has changed only in minor details since the Stone Age, and there is no chance that it can be significantly, usefully, or safely modified in the foreseeable future. This genetic permanency determines the physiological limits beyond which human life cannot be safely altered by social and technological innovations. In the final analysis, the frontiers of cultural and technological development are determined by man's genetic make-up which constitutes his own biological frontiers.

On the other hand, mankind has a large reserve of potentialities that become expressed only when circumstances are favorable. Physical and social surroundings condition both the biological and the mental expressions of individuality. Environmental planning can thus play a key role in the realization of human potentialities. One can take it for granted that there is a better chance of converting these potentialities into reality when the environment provides a variety of stimulating experiences and opportunities, especially for the young.

According to a French proverb, *Il n'y a que le provisoire qui dure* (Only that which is temporary endures). This phrase appears at first sight only a flippant expression of lazy skepticism, a denial that careful planning is worth the effort. However, it embodies a profound and universal biological truth. Living organisms can survive—whether as species or as individual specimens—only by continuously modifying some aspects of their essential being in the course of adaptive responses to the environment. Similarly, social structures can continue to prosper only by evolving. Houses grow organically through the addition of rooms, gables, and appendages to accommodate new members of the family and new social habits. The buildings and practices of churches and monasteries are modified to incorporate new interpretations of the faith and new religious attitudes. City halls at any given period reflect the problems posed by urban growth and by the multiplication of public services. Universities are presently struggling to discover how they can adapt their programs to the demands for new kinds of theoretical knowledge and for greater involvement in the practical affairs of society.

From great estate to municipal park, from slow-paced country road to multilane parkway, from city playground to national recreation area, from village to city, from suburb to satellite community, and from one-room schools to complex educational systems, the environment endlessly evolves in response to changing human needs and dreams. The concept of an optimum environment is unrealistic because it implies a static human life. Planning for the future demands an ecological attitude based on the assumption that man will continuously bring about evolutionary changes through the creative potentialities inherent in his biological nature.

The constant feedback between man and environment inevitably implies a continuous alteration of both. However, the various aspects of biological and social nature constitute such a highly integrated system that they can be altered only within a certain range. Neither physicochemical concepts of the body machine nor hopes for technological breakthroughs are of use in defining the ideal man or the proper environment unless they take into consideration the elements of the past that have become progressively incarnated in human nature and in human societies, and that determine the limitations and the potentialities of human life.

The past is not dead history; it is the living material out of which man makes himself and builds the future.

REFERENCE NOTES

Chapter 1. THE UNBELIEVABLE FUTURE

1. Lynn White, Jr., "The Historical Roots of Our Ecologic Crisis," *Science*, 155 (1967), 1203–1207.

2. Aldous Huxley, *Literature and Science* (New York: Harper, 1963), and "Achieving a Perspective on the Technological Order," in Carl F. Stover (ed.), *The Technological Order* (Detroit, Mich.: Wayne State University Press, 1963, 252–258).

3. James Reston, "Washington: The New Pessimism," *The New York Times*, April 21, 1967, 38; see also editorial, "Voices of Doubt," *The Wall Street Journal*, April 26, 1967, 16.

4. There have been many expressions of this attitude in recent scientific and sociological literature; see Stover, *op. cit.*; Jacques Ellul, *The Technological Society*, trans. John Wilkinson (New York: Knopf, 1965); Elmer Engstrom, "Science, Technology, and Statesmanship," *American Scientist*, 55 (1967), 72–79; White, *op. cit.*; also the discussion "Does Science Neglect Society?", *Science*, 158 (1967), 1134–1136.

5. John R. Platt, *The Step to Man* (New York: Wiley, 1966), 185–203.

6. Vernon Van Dyke, *Pride and Power: The Rationale of the Space Program* (Urbana, Ill.: University of Illinois Press, 1964), 155.

7. See Stover, *op. cit.*; Ellul, *op. cit.*

8. Lewis Mumford, *The Myth of the Machine* (New York: Harcourt, 1967).

9. Dean Acheson, *Morning and Noon* (Boston: Houghton Mifflin, 1965), 1.

10. Herman Kahn and Anthony Wiener, *The Year 2000, a Framework for Speculation on the Next Thirty-three Years* (New York: Macmillan, 1968).

11. Henry Adams, *The Education of Henry Adams* (Boston: Houghton Mifflin, 1906).

12. Stephen Vincent Benét, *Western Star* (New York: Farrar & Rinehart, 1943), 3, 12.

13. Pierre Teilhard de Chardin, *The Phenomenon of Man*, trans. Bernard Wall (New York: Harper, 1959); French title: *Le Phénomène Humain*).

14. Frank Lloyd Wright, *The Future of Architecture* (New York: Horizon Press, 1953), 80.

15. W. O. Roberts, "Science, A Wellspring of Our Discontent," *American Scientist*, 55 (1967), 3–14.

16. I have benefited in writing this chapter from discussions with Mrs. Julie Field, author, with Will Burtin, of a manuscript on "The Architecture of an Ethic."

Chapter 2. MAN'S NATURE AND HUMAN HISTORY

1. L. S. Cressman, *The Sandal and the Cave* (Portland, Ore.: Beaver Books, 1964).

2. *Ibid.*, 28.

3. *Ibid.*, 6.

4. See V. Gordon Childe, *Man Makes Himself* (New York: New American Library, 1961) and *Social Evolution* (New York: Meridian Books, 1963); Bernard Campbell, *Human Evolution* (Chicago, Ill.: Aldine Publishing Co., 1966); John Buettner-Janusch, *Origins of Man: Physical Anthropology* (New York: Wiley, 1966); René Dubos, *Man, Medicine and Environment* (New York: Praeger, 1968).

5. Mumford, *The Myth of the Machine, op. cit.*

6. See E. A. Speiser (ed.), *The World History of the Jewish People. Vol. I. At the Dawn of Civilization* (New Brunswick, N.J.: Rutgers University Press, 1964); Childe, *Man Makes Himself* and *Social Evolution, op. cit.*

7. Arnold Toynbee, *A Study of History* (New York: Oxford University Press, 1957), 243.

8. See Gordon Willey, *An Introduction to American Archaeology, Vol. I. North and Middle America* (Englewood Cliffs, N.J.: Prentice-Hall, 1966); Robert Lowie, *Indians of the Plains* (Garden City, N.Y.: The Natural History Press, 1963); Philip Drucker, *Indians of the Northwest Coast* (Garden City, N.Y.: The Natural History Press, 1963); Kenneth Macgowan and Joseph A. Hester, Jr., *Early Man in the New World* (New York: The Natural History Library, 1962); David Hopkins (ed.), *The Bering Land Bridge* (Stanford, Calif.: Stanford University Press, 1967); Paul Martin, "A Prehistoric Route to America" (review of *The Bering Land Bridge*, edited by David Hopkins), *Science*, 158 (1967), 1168.

9. Theodora Kroeber, *Ishi in Two Worlds: A Biography of the Last Wild Indian in North America* (Berkeley, Calif.: University of California Press, 1961).

10. Hippocrates, *Of Airs, Waters, and Places*, trans. W. Jones (New York: Putnam, 1931).

11. M. D. Coe and K. V. Flannery, "Microenvironments and Mesoamerican Prehistory," and B. J. Meggers, "Environmental Limitation on the Development of Culture," in J. B. Bresler (ed.), *Human Ecology* (Reading, Mass.: Addison-Wesley, 1966).

12. Drucker, *op. cit.*

13. Ralph Waldo Emerson, "The Uses of Great Men," *Representative Men* (Boston: Houghton Mifflin, 1883), 29.

14. See Benjamin S. Bloom, *Stability and Change in Human Characteristics* (New York: Wiley, 1964); H. Bakwin and S. McLaughlin, "Secular Increase in Height: Is the End in Sight?" *The Lancet*, II (1964), 1195–1196; A. W. Boyne, "Secular Changes in the Stature of Adults and the Growth of Children, with Special Reference to Changes in Intelligence of 11-Year-Olds," in J. M. Tanner (ed.), *Human Growth* (New York: Pergamon Press, 1960); J. M. Tanner, "The Trend Towards Earlier Physical Maturation," in J. E. Meade and A. S. Parkes (eds.), *Biological Aspects of Social*

Problems (New York: Plenum Press, 1965), and "Earlier Maturation in Man," *Scientific American*, 218 (1968), 21–27.

15. See René Dubos, *Mirage of Health* (New York: Harper, 1959) and *Man Adapting* (New Haven: Yale University Press, 1965), Chapters 7 and 9; René Dubos and Jean Dubos, *The White Plague: Tuberculosis, Man, and Society* (Boston: Little, Brown, 1952).

16. See René Dubos, *Man Adapting, op. cit.*, Chapter 11; Kingsley Davis, "Population Policy: Will Current Programs Succeed?", *Science*, 158 (1967), 730–739.

17. Leo Marx, *The Machine in the Garden: Technology and the Pastoral Ideal* (New York: Oxford University Press, 1964).

18. Theodosius Dobzhansky, *Mankind Evolving* (New Haven, Conn.: Yale University Press, 1962).

19. Speiser, *op. cit.*, 266–267.

20. José Ortega y Gasset, *History As a System* (New York: Norton, 1941), 217.

Chapter 3. BIOLOGICAL REMEMBRANCE OF THINGS PAST

1. The following general books by contemporary masters of the science of genetics present lucid and accurate statements of the problems of heredity: George W. Beadle and Muriel Beadle, *Language of Life* (New York: Doubleday Anchor Books, 1966); Francis Crick, *Of Molecules and Men* (Seattle, Wash.: University of Washington Press, 1966); Theodosius Dobzhansky, *Heredity and the Nature of Man* (New York: Harcourt, Brace and World, 1964); C. H. Waddington, *The Nature of Life* (London: Allen & Unwin, 1961).

2. See C. H. Waddington, "Evolutionary Adaptation," *Perspectives in Biology and Medicine*, II, 4 (1959), 379–401, and *The Nature of Life, op. cit.*; Alister Hardy, *The Living Stream* (New York: Harper, 1965) and "Another View of Evolution," in I. T. Ramsey (ed.), *Biology and Personality* (New York: Barnes and Noble, 1965).

3. See Herman F. Becker, "Flowers, Insects, and Evolution," *Natural History*, 74 (1965), 38–45; E. Lendell Cockrum and Bruce J. Hayward, "Hummingbird Bats: Nectar-Drinking *Leptonycteris* Is a Cactus Pollinating Agent," *Natural History*, 71 (1962), 39–43.

4. Hardy, *The Living Stream*, and "Another View of Evolution," *op. cit.*

5. See Peter Marler and M. Tamura, "Song 'Dialects' in Three Populations of White-Crowned Sparrows," *The Condor*, 64 (1962), 368–377; Peter Marler and W. Hamilton III, *Mechanisms of Animal Behavior* (New York: Wiley, 1966).

6. See J. Itani, "On the Acquisition and Propagation of a New Food Habit in the Troop of Japanese Monkeys at Takasakiyama," *Primates*, 1 (1958), 84–98; M. Kawai, "Newly-acquired Precultural Behavior of the Natural Troop of Japanese Monkeys on Koshima Island," *Primates*, 6 (1965), 1–30.

7. See Vernon Reynolds, "The 'Man of the Woods,'" *Natural History*, 73 (1964), 44–51, and *Budongo: An African Forest and Its Chimpanzees* (Garden City, N.Y.: The Natural History Press, 1965); George Schaller, *The Year of the Gorilla* (London: Collins, 1965).

8. See Clifford Geertz, "The Growth of Culture and The Evolution of Mind," in J. M. Scher (ed.), *Theories of the Mind* (New York: The Free Press, 1962), 713–740, and "The Impact of the Concept of Culture on the Concept of Man," *Bulletin of the Atomic Scientists*, 22 (1966), 2–8; Carleton S. Coon, *The Story of Man* (New York: Knopf, 1962); H. Hoagland and R. Burhoe (eds.), *Evolution and Man's Progress* (New York: Columbia University Press, 1962); S. L. Washburn (ed.), *Social Life of Early Man* (Chicago: Aldine Publishing Co., 1961); Buettner-Janusch, *op. cit.*

9. Carleton S. Coon, "An Anthropogeographic Excursion Around the World," *Human Biology*, 30 (1958), 29–42.

10. George C. Williams, *Adaptation and Natural Selection: A Critique of Some Current Evolutionary Thought* (Princeton, N.J.: Princeton University Press, 1966), 79.

11. See Alan H. Brodrick, *Man and His Ancestry* (Greenwich, Conn.: Fawcett Publications, 1964); Carleton S. Coon, "Some Problems of Human Variability and Natural Selection in Climate and Culture," *American Naturalist*, 89 (1955), 257–279, and *The Story of Man, op. cit.*; Lee R. Dice, *Man's Nature and Nature's Man* (Ann Arbor, Mich.: The University of Michigan Press, 1955); G. A. Harrison, J. S. Weiner, J. M. Tanner, and N. A. Barnicot, *Human Biology* (Oxford: Clarendon Press, 1964); S. L. Washburn (ed.), *Classification and Human Evolution* (London: Methuen, 1964); Wilfrid E. Le Gros Clark, *Man-Apes or Ape-Men?* (New York: Holt, Rinehart, and Winston, 1967).

12. See Erwin Bunning, *Physiological Clock: Endogenous*

Diurnal Rhythms and Biological Chronometry (New York: Academic Press, 1964); J. L. Cloudsley-Thompson, *Rhythmic Activity in Animal Physiology and Behavior* (New York: Academic Press, 1961); René Dubos, *Man Adapting, op. cit.*, Chapter 2.

13. See Walter Menaker and Abraham Menaker, "Lunar Periodicity in Human Reproduction: A Likely Unit of Biological Time," *American Journal of Obstetrics and Gynecology*, 77 (1959), 905–914; U. M. Cowgill, A. Bishop, R. J. Andrew, and G. E. Hutchinson, "An Apparent Lunar Periodicity in the Sexual Cycle of Certain Prosimians," *Proceedings of the National Academy of Sciences*, 48 (1962), 238–241.

14. Jack London, *The Call of the Wild* (New York: Heritage Press, 1960), 158.

15. For an unorthodox but stimulating discussion of Pavlov's work see William Sargant, *Battle for the Mind* (Baltimore: Penguin Books, 1957), 4–6.

16. J. P. Scott, "Critical Periods in Behavioral Development," *Science*, 138 (1962), 949–958.

17. General reviews of this immense and diffuse topic will be found in Urie Bronfenbrenner, "Early Deprivation in Mammals and Man," in Grant Newton (ed.), *Early Experience and Behavior* (Springfield, Ill.: Charles C Thomas, 1966); René Dubos, *Man Adapting, op. cit.*, Chapter 1; W. Sluckin, *Imprinting and Early Learning* (Chicago: Aldine Publishing Co., 1965).

18. Scott, *op. cit.*

19. Margaret Ounsted and Christopher Ounsted, "Maternal Regulation of Intra-uterine Growth," *Nature*, 212 (1966), 995–997.

20. See S. A. Barnett and J. Burn, "Early Stimulation and Maternal Behaviour," *Nature*, 213 (1967), 150–152; V. H. Denenberg, D. R. Ottinger, and M. W. Stephens, "Effects of Maternal Factors upon Growth and Behavior of the Rat," *Child Development*, 33 (1962), 65–71; S. Levine, "Stimulation in Infancy," *Scientific American*, 202 (1960), 81–86, and "Psychophysiological Effects of Infantile Stimulation," in E. L. Bliss (ed.), *Roots of Behavior* (New York: Harper, 1962), 246–253; R. Ader, "Effects of Early Experience and Differential Mothering on Behavior and Susceptibility to Gastric Erosium in the Rat," *Journal of Comparative and Physiological Psychology*, 60 (1965), 233–238.

21. Thomas K. Landauer and John Whiting, "Infantile

Stimulation and Adult Stature of Human Males," *American Anthropologist*, 66 (1964), 1007–1028.

22. See René Dubos, Russell Schaedler, and Richard Costello, "Lasting Biological Effects of Early Environmental Influences. I. Conditioning of Adult Size by Prenatal and Postnatal Nutrition," *Journal of Experimental Medicine*, 127 (1968), 783–799; R. A. McCance, "Food, Growth, and Time," *The Lancet*, II, September 29 (p. 621) and October 6, (p. 671), 1962; Nevin Scrimshaw, "Malnutrition, Learning and Behavior," *The American Journal of Clinical Nutrition*, 20 (1967), 493–502; Joaquín Cravioto, "Nutritional Deficiencies and Mental Performance in Childhood," in David Glass (ed.), *Biology and Behavior: Environmental Influences* (New York: Rockefeller University Press and Russell Sage Foundation, 1968).

23. R. C. MacKeith, "Is a Big Baby Healthy?" *Proceedings of the Nutritional Society*, 22 (1963), 128–134.

24. Harrison *et al., op. cit.*, 358–366.

25. Peter Neubauer (ed.)., *Children in Collectives—Childrearing Aims and Practices in the Kibbutz* (Springfield, Ill.: Charles C Thomas, 1965).

26. See Note 14 for Chapter 2.

27. Roy E. Brown, "Organ Weight in Malnutrition with Special Reference to Brain Weight," *Developmental Medicine and Child Neurology*, 8 (1966), 512–522.

28. Mark Rosenzweig, "Environmental Complexity, Cerebral Change, and Behavior," *American Psychologist*, 21 (1966), 321–322.

29. C. L. Pratt and G. P. Sackett, "Selection of Social Partners as a Function of Peer Contact during Rearing," *Science*, 155 (1967), 1133–1135.

30. H. F. Harlow and M. K. Harlow, "The Effect of Rearing Conditions on Behavior," *Bulletin of Menninger Clinic*, 26 (1962), 213–224; see also H. F. Harlow and M. K. Harlow, "Social Deprivation in Monkeys," *Scientific American*, 207 (1962), 136–146, and "The Affectionate Systems," in S. M. Schrier, H. F. Harlow, and F. Stollnitz (eds.), *Behavior of Nonhuman Primates*, Vol. II (New York: Academic Press, 1965), 287–334; H. F. Harlow, M. K. Harlow, and E. W. Hansen, "The Maternal Affectional System of Rhesus Monkeys," in H. F. Rheingold (ed.), *Maternal Behavior in Mammals* (New York: Wiley, 1963), 254–281.

31. See B. M. Caldwell, "The Effects of Infant Care" (9–87),

and L. J. Yarrow, "Separation from Parent During Early Childhood" (89–136), in M. L. Hoffman and L. W. Hoffman (eds.), *Review of Child Development Research* (New York: Russell Sage Foundation, 1964).

32. J. B. Calhoun, "Population Density and Social Pathology," *Scientific American*, 206 (1962), 139–148.

33. See E. T. Hall, *The Silent Language* (New York: Doubleday, 1959) and *The Hidden Dimension* (New York: Doubleday, 1966).

34. See M. F. Ashley Montagu, *Prenatal Influences* (Springfield, Ill.: Charles C Thomas, 1962); Ounsted, *op. cit.*; P. Gruenwald, H. Funakawa, S. Mitani, T. Nishimura, and S. Takeuchi, "Influence of Environmental Factors on Foetal Growth in Man," *The Lancet*, I (1967), 1026–1028.

35. Eleanor Maccoby (ed.), *The Development of Sex Differences* (London: Tavistock Publications, 1967), 13 and 49.

36. Bacon F. Chow and Roger W. Sherwin, "Fetal Parasitism?" *Archives of Environmental Health*, 10 (1965), 395–398.

37. Noam Chomsky, "Language and the Mind," *Psychology Today*, 1 (1968), 48–51 and 66–69.

38. Jean Piaget, *The Origins of Intelligence in Children* (New York: International Universities Press, 1952) and *The Construction of Reality in the Child* (New York: Basic Books, 1954).

39. See E. H. Davidson, "Hormones and Genes," *Scientific American*, 212 (1965), 36–45; S. Levine, "Sex Differences in the Brain," *Scientific American*, 214 (1966), 84–90; T. L. Campbell, "Reflections on Research and the Future of Medicine," *Science*, 153 (1966), 442–449.

40. Dr. John Brock, Professor of Medicine, University of Cape Town, South Africa, is at present preparing a book defining more completely the medical meaning of the word "constitution." In the preparation of this chapter I have greatly benefited from discussions with him.

41. A. T. W. Simeons, *Man's Presumptuous Brain* (New York: Dutton, 1961).

42. Josef Pieper, *Love and Inspiration: A Study of Plato's "Phaedrus"* (London: Faber and Faber, 1962).

43. Quoted in José Ortega y Gasset, *The Origin of Philosophy* (New York: Norton, 1967), 82.

44. E. R. Dodds, *The Greeks and The Irrational* (Berkeley, Calif.: University of California Press, 1951).

45. See George L. Engel, *Psychological Development in Health and Disease* (Philadelphia: Saunders, 1962); H. G. Wolff, *Stress and Disease* (Springfield, Ill.: Charles C Thomas, 1953), "Stressors as a Cause of Disease in Man," in J. M. Tanner (ed.), *Stress and Psychiatric Disorder* (Oxford: Blackwell, 1960), 17–33, and "The Mind-Body Relationship," in L. Bryson (ed.), *An Outline of Man's Knowledge* (New York: Doubleday, 1960), 41–72.

46. Geertz, "The Growth of Culture . . ." and "The Impact of the Concept . . .", *op. cit.*; Washburn, *Social Life of Early Man* and *Classification and Human Evolution, op. cit.*

Chapter 4. THE LIVING EXPERIENCE

1. O. R. Frisch, "Niels Bohr," *Scientific American*, 216 (1967), 145–148.

2. W. H. Auden, "The Real World," *The New Republic*, December 9, 1967, 27; see also "Ode to Terminus," *The New York Review of Books*, 11 (1968), 6.

3. Quoted in the presentation by Dr. Huston Smith (p. 19), in *The Human Mind*, 1967 Nobel Conference, Gustavus Adolphus College, St. Peter, Minnesota.

4. Siegfried Giedion, *The Eternal Present* (New York: Pantheon Books, 1962).

5. See Dodds, *op. cit.*; William Barrett, *Irrational Man: A Study in Existential Philosophy* (New York: Doubleday, 1958); Alan McGlashan, *The Savage and the Beautiful Country* (London: Chatto & Windus, 1966).

6. Speiser, *op. cit.*

7. Han Suyin, *A Many-Splendored Thing* (Boston: Little, Brown, 1953), 261.

8. S. A. Barnett, *Instinct and Intelligence* (Englewood Cliffs, N.J.: Prentice-Hall, 1967).

9. Roger Fry, *Vision and Design* (London: Chatto & Windus, 1920), 91.

10. W. H. Hudson, *A Hind in Richmond Park* (New York: Dutton, 1923).

11. "56-Hour Day 230 Feet Underground," *Medical Tribune*, March 11–12, 1967, 8.

12. Norman O. Brown, *Life Against Death* (Middletown, Conn.: Wesleyan University Press, 1959).

13. See Mircea Eliade, *The Two and the One* (London: Harvill Press, 1966), and *The Sacred and the Profane: The Nature of Religion* (New York: Harcourt, Brace & World, 1962).

14. Gaston Bachelard, *The Psychoanalysis of Fire* (Boston: Beacon Press, 1964).

15. Quoted in Maurice Friedman (ed.), *The Knowledge of Man* (New York: Harper, 1965).

16. Harvey Cox, *The Secular City* (New York: Macmillan, 1966), 42.

17. Jane Jacobs, *The Death and Life of Great American Cities* (New York: Random House, 1961), 55–56.

18. Sarvepalli Radakrishnan, *Indian Philosophy* (New York: Macmillan, 1931), I, 236; see also Radakrishnan, *The Hindu View of Life* (London: Allen & Unwin, 1956).

19. Etienne Bonnet de Condillac, *Traité des Sensations* (Paris: Librarie Hachette, 1893).

20. F. A. Lange, quoted in H. Vaihinger, *The Philosophy of "As If"* (London: Kegan Paul, Trench, Trubner, 1924), 192.

21. L. K. Frank, *On the Importance of Infancy* (New York: Random House, 1966).

22. Piaget, *The Origins of Intelligence in Children* and *The Construction of Reality in the Child, op. cit.*

23. J. Z. Young, *Doubt and Certainty in Science* (New York: Oxford University Press, 1960), 36.

24. Philippe Aries, *Centuries of Childhood: A Social History of Family Life* (New York: Knopf, 1962).

25. See Philip Solomon *et al.* (eds.), *Sensory Deprivation* (Cambridge, Mass.: Harvard University Press, 1961); D. W. Fiske and S. R. Maddi, *Functions of Varied Experience* (Homewood, Ill.: Dorsey Press, 1961).

26. J. Z. Young, *A Model of the Brain* (Oxford: Clarendon Press, 1964).

27. See Sargant, *op. cit.*; Eric Salzen and C. C. Meyer, "Imprinting: Reversal of a Preference Established during the Critical Period," *Nature*, 215 (1967), 785–786.

28. See Eugene P. Wigner, "Explaining Consciousness," *Science*, 156 (1967), 798–799, and C. H. Waddington, "No Vitalism for Crick," *Nature*, 216 (1967), 202–203; both articles are reviews of Francis Crick, *Of Molecules and Men* (Seattle: University of Washington Press, 1966).

29. Donald MacKay, *Freedom of Action in a Mechanistic Universe* (Cambridge: Cambridge University Press, 1967).

30. Frisch, *op. cit.*

31. Jacob Bronowski, *The Identity of Man* (Garden City, N.Y.: The American Museum of Natural History, 1965), 2–10.

32. George Wald, "Determinancy, Individuality, and the Problem of Free Will," in John R. Platt (ed.), *New Views of the Nature of Man* (Chicago: University of Chicago Press, 1965), 40.

33. Corneille J. F. Heymans, in Giulio Gabbiani (ed.), *Reflections on Biologic Research* (St. Louis, Mo.: Warren H. Green, Inc., 1967), 84; see also Keller Breland and Marian Breland, "The Misbehavior of Organisms," in Thomas E. McGill (ed.), *Readings in Animal Behavior* (New York: Holt, Rinehart and Winston, 1965), 455–460.

34. José Ortega y Gasset, *The Dehumanization of Art and Other Writings on Art and Culture* (Garden City, N.Y.: Doubleday Anchor Books, 1956), 153.

35. Feodor Dostoevski, *Notes from Underground,* in *The Short Novels of Dostoevsky,* trans. Constance Garnett (New York: Dial Press, 1945), 145, 149.

Chapter 5. THE PURSUIT OF SIGNIFICANCE

1. Quoted in John M. Rich, *Chief Seattle's Unanswered Challenge* (Seattle: John M. Rich, 1932), 30–31.

2. *Ibid.*, 33, 36, 40.

3. Robert W. Young and William Morgan, *Navajo Historical Selections,* Navajo Historical Series No. 3, Bureau of Indian Affairs, 1954.

4. Rabindranath Tagore, *Towards Universal Man* (New York: Asia Publishing House, 1961), 294.

5. Rich, *op. cit.*, 36.

6. George Homans, *English Villagers of the Thirteenth Century* (New York: Russell & Russell, 1960).

7. W. C. Loring, comments on "City Planning and the Treasury of Science" by John W. Dyckman, in William R. Ewald, Jr. (ed.), *Environment for Man* (Bloomington, Ind.: Indiana University Press, 1967), 52–56.

8. Quoted in Stanley M. Garn, *Culture and the Direction of Human Evolution* (Detroit, Mich.: Wayne State University Press, 1964), 16; see also Charles P. Mountford, *Ayers Rock: Its People, Their Beliefs, and Their Art* (Honolulu: East-West Center Press, 1965).

9. Vilhjalmur Stefansson, *The Friendly Arctic* (New York: Macmillan, 1953).

10. René Dubos, *Man Adapting, op. cit.*, Chapter 10.

11. Dickinson W. Richards, "Homeostasis: Its Dislocations and Perturbations," *Perspectives in Biology and Medicine*, 3 (1960), 238–251.

12. René Dubos, *Man Adapting, op. cit.*, Chapter 9.

13. See Richard Neutra, *Survival Through Design* (New York: Oxford University Press, 1954); Christopher Alexander, "The City as a Mechanism for Sustaining Human Contact" (60–102), and Moshe Safdie, "Habitat '67" (253–260), in Ewald, *op. cit.*

14. Alexander, *op. cit.*, 74–86.

15. Bachelard, *op. cit.*

16. D. H. Lawrence, *Studies in Classic American Literature* (New York: Thomas Seltzer, 1923), 201, 202, 203, 206.

17. William L. C. Wheaton, "Form and Structure of the Metropolitan Area," in Ewald, *op. cit.*, 157–184.

18. Toynbee, *op. cit.*

19. Ellsworth Huntington, *Civilization and Climate* (New Haven: Yale University Press, 1924).

20. René Dubos, *Mirage of Health, op. cit.*, and *Man Adapting, op. cit.*, Chapters 7 and 9.

21. Quoted by Alexander in Ewald, *op. cit.*, 64.

22. René Dubos, *Man Adapting, op. cit.*, Chapter 3.

23. Childe, *Social Evolution, op. cit.*, 21.

24. Winston Churchill, *Onwards to Victory: War Speeches by the Right Honorable Winston Churchill*, Charles Eade (ed.) (Boston: Little Brown, 1944), 316–318.

25. Neubauer, *op. cit.*

26. Allon Schoener (ed.), *The Lower East Side: Portal to American Life* (New York: The Jewish Museum, 1966).

27. Mumford, *The Myth of the Machine, op. cit.*, 76.

28. Lewis Mumford, *Sticks and Stones* (New York: Dover Publications, 1924).

29. Wolfgang Braunfels, "Institutions and Their Corre-

sponding Ideals," in *The Quality of Man's Environment* (Washington, D.C.: Smithsonian Institution Press, 1968).

30. Philip Johnson, "Why We Want Our Cities Ugly," in *The Quality of Man's Environment, op. cit.*

31. Norbert Wiener, *The Human Use of Human Beings: Cybernetics and Society* (Garden City, N.Y.: Doubleday, 1950).

32. Henry Adams, *Mont St. Michel and Chartres* (Boston: Houghton Mifflin, 1930).

33. See Barry Commoner, *Science and Survival* (New York: Viking, 1966); Paul Sears, *The Living Landscape* (New York: Basic Books, 1967) and *Deserts on the March* (Norman, Okla.: University of Oklahoma Press, 1959); LaMont C. Cole, "Man's Ecosystem," *BioScience,* 16 (1966), 243–248; Lewis Herber, *Our Synthetic Environment* (New York: Knopf, 1962).

34. René Dubos, *Man Adapting, op. cit.*, Chapter 9.

35. See N. E. Himes, *Medical History of Contraception* (New York: Gamut Press, 1963); B. E. Finch and H. Green, *Conception through the Ages* (London: Peter Owen, 1963).

36. William L. Thomas, Jr. (ed.), *Man's Role in Changing the Face of the Earth* (Chicago: University of Chicago Press, 1956), 70.

37. Elmer W. Engstrom, "Science, Technology, and Statesmanship," *American Scientist,* 55 (1967), 72–79.

38. Stewart L. Udall, "Can America Outgrow Its Growth Myth?" Address before the Long Island Conference on Natural Beauty, Hofstra University, Spring, 1966.

39. Ritchie Calder, "The Tyranny of the Expert," paper presented at a symposium on The Technological Society, Center for the Study of Democratic Institutions, December 1965.

Chapter 6. THE SCIENCE OF HUMANITY

1. Harvey G. Cox, "Technology and Democracy," in *Technology and Culture in Perspective* (Cambridge, Mass.: The Church Society for College Work, 1967), 12.

2. Rabindranath Tagore, *op. cit.*, 294.

3. Antoine de Saint Exupéry, *Le Petit Prince* (New York: Reynal & Hitchcock, 1943)

4. Antoine de Saint Exupéry, *The Little Prince* (New York: Reynal & Hitchcock, 1943).

5. Marcus Tullius Cicero, *De natura deorum*, trans. H. Rackham (New York: Putnam, 1933), 271.

6. David Lowenthal and Hugh C. Prince, "The English Landscape," *The Geographical Review*, 54 (1964), 309–346.

7. Sean Jennet, "Britannia Deserta," *Landscape*, 15 (1965–66), 22–29.

8. Marx, *The Machine in the Garden, op. cit.*

9. Jacquetta Hawkes, *A Land* (New York: Random House, 1951), 143.

10. See Mumford, *Sticks and Stones, op. cit.*, and *The City in History* (New York: Harcourt, Brace & World, 1961); Edmund Bacon, *The Design of Cities* (New York: Viking, 1967).

11. Quoted in William L. Thomas, Jr. (ed.), *Man's Role in Changing the Face of the Earth, op. cit.*, 1112.

12. Anne Morrow Lindbergh, "Immersion in Life: Journey to East Africa," *Life*, October 21, 1966, 88.

13. Giedion, *op. cit.*

14. Quoted in Sibyl Moholy-Nagy, "The Four Environments of Man," *Landscape*, 16 (1966–67), 4.

15. Ray Bradbury, *Fahrenheit 451* (New York: Ballantine Books, 1953).

16. Stewart L. Udall, "Our Perilous Population Implosion," *Saturday Review*, September 2, 1967.

17. National Academy of Sciences, *The Scientific Endeavor* (New York: The Rockefeller Institute Press, 1965).

18. Stephen Toulmin, *Foresight and Understanding* (Bloomington, Ind.: Indiana University Press, 1961), 10.

19. Norman O. Brown, *op. cit.*

20. Speiser, *op. cit.*, 150.

21. Quoted in E. Cassirer *et al.*, *The Renaissance Philosophy of Man* (Chicago: University of Chicago Press, 1948), 224–225.

22. Childe, *Man Makes Himself, op. cit.*, 188.

23. See Hermann J. Muller, "The Guidance of Human Evolution," *Perspectives in Biology and Medicine*, 3 (1959), 1–43, and "Genetic Progress by Voluntarily Conducted Germinal Choice," in Gordon Wolstenholme (ed.), *Man and His Future* (Boston: Little, Brown, 1963), 247–262; Joshua Lederberg, "Biological Future of Man," in Wolstenholme, *op. cit.*, 263–274, "Experimental Genetics and Human Evolution," *Bulletin of the Atomic Scientists*, October 1966, 4–11, and reply to L. Ornstein, "The Population Explosion,"

ibid., June 1967, 60–61; Kingsley Davis, "Sociological Aspects of Genetic Control," in John D. Roslansky, *Genetics and the Future of Man* (Amsterdam: North Holland Publishing Co., 1966), 171–204.

For a critical discussion of these views, see P. B. Medawar, *The Future of Man* (New York: New American Library, 1959); Dobzhansky, *Mankind Evolving, op. cit.*; René Dubos, *Man Adapting, op. cit.*, Chapter 11; Curt Stern, "Genes and People," *Perspectives in Biology and Medicine,* 10 (1967), 500; T. M. Sonneborn (ed.), *The Control of Human Heredity and Evolution* (New York: Macmillan, 1965).

24. Max von Pettenkofer, *The Value of Health to a City*; two popular lectures delivered on March 26 and 29, 1873, in the *Verein fur Volksbildung* in Munich, quoted in *Bulletin of the History of Medicine,* 10 (1941), 487–503; Benjamin Richardson, *Hygeia: A City of Health* (London: Macmillan, 1876).

25. Ellul, *op. cit.* See also Stoner, *op. cit.*; John K. Galbraith, *The New Industrial State* (Boston: Houghton Mifflin, 1967).

26. Oscar Wilde, "The Soul of Man under Socialism," in *The Works of Oscar Wilde* (New York: Lamb Publishing Co., 1909), VIII, 148.

27. Charles Abrams, *The City Is the Frontier* (New York: Harper, 1965), 388.

28. Adlai E. Stevenson, speech given before the Economic and Social Council, Geneva, Switzerland, July 9, 1965.

29. Commoner, *op. cit.*

30. Bacon, *op. cit.*

31. Quoted in John W. Reps, *The Making of Urban America: A History of City Planning* (Princeton, N.J.: Princeton University Press, 1965), 344.

32. Bradbury, *op. cit.*, 8.

INDEX